D0996562

WAR
IN THE
STREETS

WAR
IN THE
STREETS

The Story of Urban Combat
from Calais to Khafji

COLONEL MICHAEL DEWAR

BCA
LONDON · NEW YORK · SYDNEY · TORONTO

To
Lavinia, Alexander, James,
Edward and Katharine
with love

Frontispiece: A British soldier equipped as he might be
in an urban environment in NBC conditions. He is
wearing the new British Army helmet, a Mk 3
respirator, an NBC suit and is carrying the new Royal
Ordnance SA80 rifle

This edition published 1992 by BCA by arrangement with
DAVID & CHARLES PLC

Book designed by Michael Head.

Typeset by Ace Filmsetting Ltd, Frome, Somerset
and printed in Great Britain by BPCC Hazell Books, Paulton & Aylesbury

Contents

A British infantryman at the ready in an urban setting. He is armed with a 7.62mm SLR, recently replaced by the 5.56mm SA80. This photograph was taken on exercise Brave Defender in the UK in 1985

Introduction

The British Army calls it 'Fighting in Built-up Areas', or FIBUA. The US Army calls it 'Military Operations in Urban Terrain', or MOUT. It has other names, too, such as 'conventional warfare in an urban environment'. Most straightforwardly of all it is sometimes called 'street fighting'. Whatever we choose to call it, urban combat—which is the term used throughout this book—is becoming increasingly relevant in modern warfare. Vast sums of money are being invested by NATO armies in building urban combat training facilities. The British Army alone has one major facility completed in West Berlin, a second more recently completed on Salisbury Plain in the United Kingdom, and a third planned at Sennelager in Germany. Why have military men suddenly rediscovered urban combat?

Battles have, of course, been fought in towns and villages for a long time, though mostly during and since World War II; by which time the pace of urbanisation had increased dramatically. In fact the urbanisation of Western Europe has now reached a stage where in any future war it would be difficult *not* to fight in urban areas, and urban combat has consequently become a technique of major military significance. However, although much has been written about it in history, particularly during World War II, little of significance has been written on contemporary theory and practice.

It would be impossible to tackle this subject without setting it in an historical context, but it is the contemporary scene with which this book is mainly concerned. Since the revolution of 1989–90 in Europe it is the common expectation that the incidence of warfare is likely to be reduced. It is of course true that the likelihood of war

between the Soviet Union and NATO has been vastly reduced, particularly since the signing of the Conventional Forces in Europe (CFE) Treaty in November 1990. However, the new freedoms have given a new impetus to traditional rivalries and old enmities. In 1991 tensions exist between Hungary and Romania concerning the Hungarian majority in Romania; Moldavians and Romanians claim a common ethnic heritage; there is a substantial Turkish minority in Greece; Cyprus remains the main impediment to normal relations between Greece and Turkey; the Greek minority in Albania is unhappy; the Romanian revolution is seemingly still in progress; Yugoslavia remains in a state of flux; tensions between Czechs and Slovaks show no signs of going away; the Poles are still suspicious of a united Germany; the Baltic States are insistent that they should gain their independence; many of the Soviet republics, particularly in the south, are demanding a degree of independence. All these examples are in Central or Eastern Europe, and in relatively urbanised societies. But potential urban unrest is unlikely to be confined to Europe. There are situations the world over which could easily result in troops being used in urban situations—most recently, we have witnessed vicious urban combat in the streets of Khafji, Kuwait, in February 1991.

Of all military techniques, urban combat is undoubtedly the one above all others with which military forces worldwide must remain constantly up-to-date. So, whilst highlighting relevant lessons from the past, I have at the same time tried to document contemporary doctrine and practice as well as look ahead to possible future developments. Unfortunately, urban combat is likely to be the most persistent of all forms of warfare.

1   The Nature of Urban Combat

'What is the position about London? I have a very clear view that we should fight every inch of it, and that it would *devour* quite a large invading army.' These memorable words were contained in a memorandum from Winston Churchill to General Ismay, penned on 2 July 1940, when an invasion of the United Kingdom by Hitler's armies—poised to cross the Channel from the ports on the northern coast of France—was a real possibility. Churchill, probably before anyone else, appreciated the defensive potential of a large conurba-

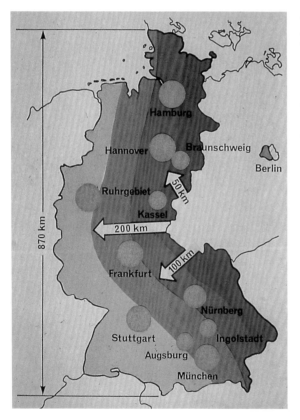

Restricted space and major conurbations in West Germany

tion. London's potential was not, in the event, put to the test, but that of Cassino, Stalingrad and Berlin was—as indeed was that of scores of other cities in Europe in the years 1940–5. If so much urban combat took place during World War II, how much greater would its incidence be in any future conflict in an increasingly urbanised world?

If one compares a map of north-west Europe today to one printed twenty-five years ago it would be difficult to recognise it as covering the same area. The degree of urbanisation has been dramatic. This is particularly the case in what was formerly the Federal Republic of West Germany, whose towns and cities were bombed relentlessly by the RAF and USAF from 1943 onwards, and fought over by the wartime allies in 1944–5, causing immense destruction and damage. The German nation, and its economy, was then in ruins; but like a Phoenix rising from the ashes, it revived, and boomed for forty years. The result of that boom has been urban development on an unprecedented scale—the phenomenon of urban sprawl is becoming more and more apparent, where villages, towns and cities are relentlessly expanding their boundaries and merging one into the other.

There is a wealth of evidence, too, which suggests that, despite forecasts of a demographic trough in the 1990s, the gradual urbanisation of Germany is likely to continue. It is already the case that some 20 per cent of the surface area of the former Federal Republic is urbanised. This in itself is sufficient reason for a renewed interest amongst European military powers in FIBUA. But there are other good reasons. The first of these is that it is safer to plan on the assumption that an invading Soviet army will undoubtedly fight through an urban area if it offers them advantages. For instance, a built-up area providing a communication centre for road, rail or power links might well be assaulted, as might a heavily defended area which offered a chance for enemy formations to be encircled or destroyed. Both would be equally important as military targets. Although the revolutions of 1989–90 and the subsequent Soviet peace offensive has made the likelihood of armed conflict in Central Europe vastly less likely, military strategists must still plan for the possibility of conflict—not so much in the context of NATO and the Soviet Union, but within Central or Eastern Europe or even *inside* the Soviet Union.

The second reason for urban combat assuming a higher profile is that modern methods in building design and construction are unlikely to produce the same sort of constricting effects as was the case in Stalingrad, Goch or Berlin. Modern cities include wide, straight boulevards and urban motorways. Manoeuvre is possible. Housing densities are less. Buildings nowadays are built for the most part using reinforced concrete, and are therefore unlikely to produce as much rubble as their masonry predecessors. Weapons systems, too, have changed, and there are now a variety of systems which make FIBUA a less daunting prospect. Finally, history reveals that immobile forces in built-up areas can be bypassed or destroyed by artillery fire—but a defence based on strong points and manoeuvrable forces is a much more difficult nut to crack. So there are many reasons why FIBUA is becoming increasingly important.

In the 1960s, 1970s and up to the mid-1980s interest in FIBUA waned, probably because it was perceived that Soviet formations would try to avoid urban fighting and its resultant delays; and also because towns tend to soak up large numbers of defending troops. The problems of collateral damage and of operating in severely damaged cities were also unattractive to Western armies. But perceptions have gone full circle; the stark fact is that Soviet forces, or any other attacking armies, may no longer have the choice. They will be faced by vast conurbations and they will have to fight through them whether they like it or not. Moreover it may well be to the advantage of NATO, or any other defending forces, to make the most of this changing balance, and to hinge part of their defences around built-up areas.

Urban combat covers a spectrum of situations, techniques and skills, best considered under headings of offensive and defensive operations.

Offensive Operations

In the *offence*, there are three ways of dealing with a built-up area: it can be bypassed, it can be neutralised (by artillery and air bombardment), or it can be attacked. The leading formations in an advance may be ordered to bypass a defended urban area so as to maintain the momentum of the advance, and also to induce the defenders to withdraw by threatening their rear. The Soviets, as a result of their experiences in World War II, are particularly aware of the dangers of loss of impetus, fragmentation of forces and of loss of centralised control—all of which are likely to happen if an army gets bogged down in an urban area. Current Soviet policy therefore is to bypass defended towns if at all possible, whilst deploying a flank guard to protect their forces on their main axis and leaving it to second echelon formations to reduce urban garrisons. Those towns or cities, however, which the Soviets feel *should* be taken due to their political, strategic or tactical importance, would be assaulted after detailed planning and preparation.

NATO troops are unlikely to be involved in offensive urban combat in a European war. Nevertheless, history has demonstrated that all armies need to retain this capability—it may not always be possible to follow either the bypass or the neutralisation options, so NATO troops must be capable of mounting a deliberate attack to clear an urban area systematically. This is a difficult, lengthy and costly operation which would require at least a battalion to achieve. The complexity of routes, opportunities for concealment, and the variety of approaches offered by urban areas, all means that a commander must acquire as much information as he possibly can concerning city layout and construction, as well as the latest enemy dispositions, if he is to come up with a workable plan of attack.

Whatever his detailed plan, experience has shown that it should be divided into three phases: isolation of the area; the assault; and clearance.

Isolation

The aim is to isolate the area by seizing those features which dominate the approaches to the city. Enemy defences and the terrain may prevent complete isolation. The minimum requirement is to

(Opposite, above) A typical wide European city street, affording easy access for armoured vehicles. This sort of wide boulevard would be difficult to block

(Opposite, below) An aerial view of Hamburg. It is easier to progress through this sort of urban landscape than the much more congested streets of pre-1945 Europe

(Above) This scene illustrates the close quarter nature of urban combat. Targets are a house away: petrol bombs may be hurled from the roof above: obstacles are numerous

secure positions outside the town from which it is possible to give fire support to troops gaining a foothold in the town. If it is possible to hamper the enemy's withdrawal and prevent reinforcement and resupply, so much the better.

Assault
Having achieved whatever degree of isolation is possible, the commander should initiate the assault without delay. This will usually involve a deliberate attack to gain a foothold in the city, and is best launched at night or under cover of smoke. This may well be the most difficult phase of the attack since, if the enemy has chosen his ground well, he will be in a position to dominate the approaches to the city limits, both by observation and with fire. He will probably be under cover; those attacking are more likely to be in the open. Artillery and tank support will almost certainly be required to gain the overwhelming degree of firepower necessary to achieve a lodgement. However, once a break-in has been achieved, the attacker is at least now fighting in the same dimension as the enemy, however difficult the subsequent street-fighting and house-clearing may be.

Clearance
The next stage is to clear the town. Seizing key objectives will involve the attacking troops working their way through the urban landscape by a variety of routes. It may be possible for them to use streets by avoiding the open areas exposed to enemy fire, but it is more likely that they will have to resort to those gardens and backyards parallel to the streets, the roofs of buildings, the interiors of houses (by 'mouse-holing' through dividing walls), and even to underground approaches using sewers and drains. Progress will be slow, sometimes painfully slow, as they fight their way from house to house. The clearance phase is characterised by a series of squad actions, all of which are designed to accomplish the methodical clearance of particular zones. In extended built-up areas it may sometimes be necessary to clear just a corridor as a means of crossing the area.

Street-clearing and house-clearing require special skills, as fighting will be at very close quarters. The enemy will be a house or a street away, sometimes only a room away or the other side of the

wall, so the infantryman in urban combat has to be prepared for hand-to-hand fighting. His reactions will have to be instantaneous, there will be no second chances. Another very real problem is the difficulty of locating the source of enemy fire in a built-up area. The 'crack' and 'thump' of a high velocity round echoes and re-echoes off surrounding buildings so that it is virtually impossible to be sure from which direction it has been fired. This is a phenomenon which British soldiers face constantly in the streets of Belfast. The situation will be further confused by the smoke and dust which will inevitably hang in the streets. Fields of fire and of observation are much more restricted than the infantryman is used to, although there is probably more and better cover than in a rural environment. It is invariably the attacker who must expose himself in order to make progress. Snipers can, of course, make life very difficult and are usually so well hidden that it is difficult to get an aimed shot at them. The best answer is to fire an anti-tank weapon at the window or part of the house from which it appears the fire is coming. Overkill perhaps, but effective nonetheless.

Urban combat is a peculiarly infantry skill. This is not to say that tanks cannot provide very effective close support, but they must be closely protected by infantry. If a tank ventures along an enemy-held street without the houses on either side being cleared first, an anti-tank weapon can be fired into the side or rear of the tank at very close quarters. Tanks *can* be a battle-winning weapon in urban combat, but they must be carefully used; really they are designed for use in open country where they can survive by moving fast and killing the enemy by taking him on at long range.

However, although NATO troops *could* be involved in offensive urban combat in some 'brush fire' war, their most likely role in an urban setting in Europe is a defensive one.

Defensive Operations
The business of fighting from a building needs careful thought and a great deal of preparation. To be a strongpoint, it is important that the right building is selected: if it is too small, then a single hit from an artillery shell or a direct hit from a tank could well kill everyone in the building. On the

other hand, a building that is too large may well result in the defenders being spread so thinly that they will be unable either to cover all the approaches, or provide an adequate concentration of fire to prevent the enemy storming the building.

The type of construction is also important. Old farmhouses and village houses in Germany are often timber-framed with a daub or brick infill; they are inflammable and easily reduced to rubble, particularly with a tank main armament. Modern bungalows or small two-storey houses, often prefabricated from light materials (including heavy plywood) and constructed from light brick, do not even provide protection from small arms fire. High-rise buildings are in danger of progressive collapse if they are damaged on the lower floors.

There is little doubt that the more traditional masonry house with strong walls made of brick and stone and probably consisting of three or four floors is much more suitable for defence. This type of house is usually pre-1940, has smaller windows, is much less inflammable and has—certainly in Europe—good solid cellars. There are various ways in which it is possible to strengthen a chosen strongpoint: sandbags against walls, cupboards or chests-of-drawers or even mattresses filled with earth, and staircases and doorways barricaded.

Movement around the building should be via holes in walls and ceilings through which the defenders can come and go by means of ladders, ropes and even piled-up furniture. Ceilings can be prepared for the shock of explosions by being shored up with strong timbers resting on a solid base and wedged into solid parts of the building.

Having prepared the building, the next task is to site weapons appropriately. Automatic weapons should generally be sited near ground level: a machine gun covers a long beaten zone if the rounds are travelling parallel to the ground—its potential coverage must be greater than if the gun is sited firing downwards. Snipers, on the other hand, who are engaging lone targets, are better sited up high, where they can see further; though a rifleman firing through a window should be as far back from it as possible so that he cannot be seen from the outside. Even better, he should construct a loophole in an unexpected place such as underneath a window sill or through the tiles in a roof. Hand-held anti-tank weapons are generally better sited in upper storeys so that they can fire downwards onto the less-well-protected tops of armoured vehicles. However, these weapons do have a considerable backblast and must therefore be fired from a large room and well away from the opposite wall.

But urban sprawl does leave gaps. Much of the north German plain between Hanover and the river Weser, for instance, is characterised by a mosaic of evenly dispersed but well-defined small villages, 1–3 kilometres apart, with relatively flat countryside consisting of waving cornfields in between. The effect is of a latticework, with the view from one village to the next in each direction usually uninterrupted. If each of these villages is turned into a small fortress manned by, say, an infantry company supported by a troop of tanks (in modern military parlance, a 'combat team'), a degree of enemy penetration can be accepted—he is simply bounced from 'pin' to 'pin' on the 'pintable'.

So urban combat is not just about fighting within built-up areas, it also includes fighting from and between villages. In a nutshell, whereas urban combat was to an extent the exception in World War II, it has now become an integral part of conventional warfare. And this is not just within the context of warfare in north-west Europe. One has only to look at the street battles of Hué during the Vietnam war, or the gunbattles in the streets of Algiers, Aden or Nicosia in the 1950s and 1960s; or more recently at the vicious fighting in the streets of Beirut, or the terrorist activity in Belfast or Londonderry, to realise that urban combat has spread increasingly across the whole spectrum of armed conflict.

Although modern weaponry has changed many of the techniques of urban combat, much also has remained the same. Thus it is instructive to look at modern techniques and current trends within their historical context.

This type of solid brick building provides the best protection for soldiers putting a built-up area into a state of defence. They are strongly built, probably have cellars and can withstand small arms fire

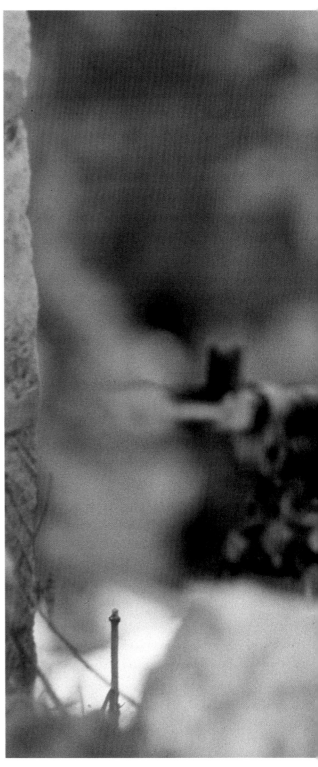

Urban combat primarily involves the infantry soldier. Here a machine gunner takes on a target a street away

2   Urban Combat in World War II: 1940–2

Armies have fought in towns and villages throughout history. Decisive actions, however, were virtually always in open countryside, and fighting actually in a town was usually confined to its subsequent sacking after its defenders had been defeated in open battle. If a fortified town was besieged, either the occupants were starved until surrender, or they sallied forth to confront their tormentors on the field of battle outside the city walls. There was no need for urban combat as there was plenty of room to fight elsewhere.

World War I, the first large-scale post-Industrial Revolution war, also saw surprisingly little urban combat. There was fighting in and around villages, and the holding of some towns was crucial to the outcome of a battle—a good example was Verdun. Pétain said that the Germans would not pass—'Ils ne passeront pas'—and nine months and 500,000 dead later they still had not passed. Verdun provided the focal point, a centre of communications, a fortress; it is doubtful if such a defence could have been organised in open countryside. But even Verdun cannot be described as urban combat in its real sense: there was no fighting in the streets, and it was merely the focus for the prolonged defence of a vital salient in the Western Front. And of course it was symbolic of the will of France to fight and win.

The Spanish Civil War was the first conflict in which urban combat became a regular phenomenon. There were battles in many towns all over Spain, but the first prolonged street fighting took place in Madrid in the autumn of 1936. The Republican forces decided to make a stand in Madrid, taking up Pétain's Verdun watchword: (in Spanish) 'No Pasaran!' ('they shall not pass'). The main battle took place between 8 and 18 November when the International Brigades, consisting mostly of German, French and Italian veterans of World War I and of victims of Fascist concentration camps, turned the tide by defeating the Nationalist effort to penetrate the defences of Madrid via the university. There was some exceptionally vicious street fighting.

Madrid was not the only example. Earlier the same year Nationalist forces had been holed up for two months in the Alcazar in Toledo, and during August and September there was savage fighting in the narrow, winding streets of this medieval city. Then right at the end of the war, in March and April 1939, the Republicans barricaded the streets of Barcelona. And death came to the city streets of Spain from the air, too. One of the most notorious examples of the use of airpower against a civilian population took place on 26 April 1937 when the fliers of the German Condor Legion bombed the town of Guernica in the Basque region, killing 1,600 and wounding 900 civilians.

But it was not until World War II that the urban landscape really became a battlefield.

Calais
By late May 1940, after the extraordinary advance of the Wehrmacht through France during the preceding weeks, the British Expeditionary Force was concentrated around Dunkirk and Calais. The decision had already been taken to evacuate the bulk of the BEF from Dunkirk when at 0300 hours on 24 May the War Office ordered the evacuation of Calais 'in principle'. The 30th Infantry Brigade occupied a perimeter around Calais about six or seven miles in length. The brigade was about 3,000 men strong and consisted of two infantry battalions, the 60th Rifles and the Rifle Brigade, and the 3rd Royal Tank Regiment and 229 Anti-Tank Battery, Royal Artillery. They had been reinforced by the 1st Batallion, Queen Victoria's Rifles on 22 May, a Territorial Army battalion which had been hastily dispatched from London to bolster the defences of Calais. The brigade was commanded by Brigadier Claude Nicholson.

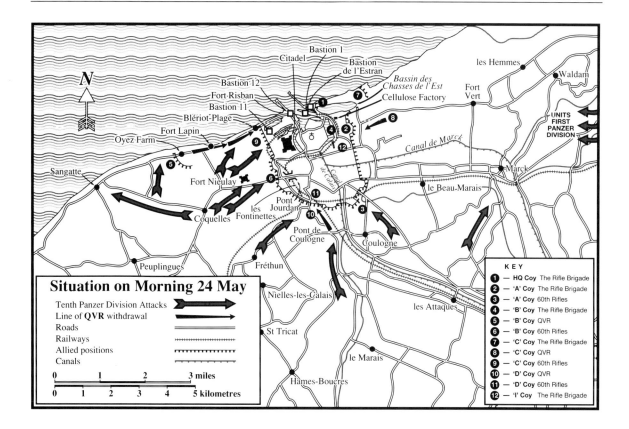

Situation on Morning 24 May

Tenth Panzer Division Attacks
Line of **QVR** withdrawal
Roads
Railways
Allied positions
Canals

0   1   2   3 miles

0   1   2   3   4   5 kilometres

K E Y
❶ — **HQ Coy** The Rifle Brigade
❷ — **'A' Coy** The Rifle Brigade
❸ — **'A' Coy** 60th Rifles
❹ — **'B' Coy** The Rifle Brigade
❺ — **'B' Coy** QVR
❻ — **'B' Coy** 60th Rifles
❼ — **'C' Coy** The Rifle Brigade
❽ — **'C' Coy** QVR
❾ — **'C' Coy** 60th Rifles
❿ — **'D' Coy** QVR
⓫ — **'D' Coy** 60th Rifles
⓬ — **'I' Coy** The Rifle Brigade

General Guderian, commanding the German Panzer divisions in the Pas de Calais area, ordered the 10th Panzer Division under General Schaal to take Calais. His division consisted of the 86th German Rifle Regiment and the 90th Tank Regiment, the latter comprising two tank battalions each with a hundred tanks and as many other fighting vehicles.

French forces in Calais were placed under the command of Nicholson by order of General Fagalde of the 16th French Corps at Dunkirk, under whose overall command the Calais garrison had now come. Hitherto there had been little cooperation between the French and British forces. However, the senior French officer in Calais, Commandant Le Tellier, rallied a total of about 800 men in the vicinity of the citadel, and they fought to the end. It also has to be said that there were in addition several thousand French and Belgian service-men who waited passively in cellars for the town to fall.

By the 24 May the Germans had closed in on

Calais: 24 May

Calais from all directions and there was heavy fighting in the suburbs. A graphic description is provided by Airey Neave in his *Flames of Calais*:

'A steady hail of tracer bullets and some tank shells came flying over the hump of the Pont Jourdan railway bridge. They bounced off the paving stones in all directions as I clung for life to the walls of houses on the south side of the boulevard and crept towards the bridge. This was my first experience of street fighting and I was acutely frightened . . . Throughout the battle, the noise was so great that if you were more than ten yards away it was impossible to understand what was said to you.'[1]

The 25 May was the crucial day of the battle. German and British troops were holding positions only 200 yards apart. Despite the aid of artillery and Stuka dive-bombers, successive German attacks were repulsed. The tenacity and courage of the British staggered General Schaal and his staff in Headquarters 10 Panzer Division. By this stage the riflemen of the 60th Rifles had forcibly occupied many of the small terraced houses in Calais-Nord. They had been unable to get into them from the front because they were in full view of the enemy, so work had at once begun on breaking into them from the back, and through the side walls for the purpose of communication. But they were not properly equipped for the job, and although work went on for nearly thirty-six hours, they never completed the task. They did not know it, but they had pioneered a technique which a year later was to become known as 'mouse-holing' in the street-fighting schools that were set up by the British Army in England and Scotland to prepare for the expected invasion across the Channel.

The street fighting in Calais was essentially haphazard, since at this time proper techniques of street fighting had not been developed. The importance of Calais, however, is that it alerted the British Army to the realities of urban combat—in this respect it was truly a baptism of fire. The scene on the 26 May was one of chaos and desolation. A death struggle had developed on the bridges leading to the harbour; the houses in the area had long since been reduced to heaps of rubble, and the surviving British officers and riflemen were left manning barricades of burned-out trucks. Consequently they had little protection from the Stukas. By this stage all hope of evacuation had faded; though the previous day Winston Churchill had sent the following message:

> 'Every hour you continue to exist is of the greatest help . . . You must continue to fight . . . have the greatest admiration for your splendid stand.'[2]

But by the afternoon of Saturday 26 May the outcome was no longer in any doubt, despite the many individual acts of defiance as the men of the Rifle Brigade, who were now pinned with their backs against the sea in the area of the Gare Maritime, made their last stand. At 3pm the Germans entered the citadel and Brigadier Nicholson was captured. Desultory resistance continued for a

few hours. Then at 7.50pm the following message was transmitted from Dover:

To OC Troop Calais
From Secretary of State

Am filled with admiration for your magnificent fight which is worthy of the highest tradition of the British Army.[3]

There was no reply.

Whether the sacrifice of Nicholson's brigade helped to gain time for the successful evacuation of the bulk of the BEF from Dunkirk is a question of endless debate. One thing is certain: the relevance of developing an expertise in urban combat became immediately apparent. For that reason alone their sacrifice was certainly not in vain.

Stalingrad
No account of urban combat during World War II can omit mention of Stalingrad. But the battle for the city need never have happened. In the spring of 1942, Stalingrad was on the northern flank of the German push into the Caucasus, the aim of which was to capture the Russian oilfields. Hitler's Directive No 41 of 5 April 1942 stated that:

> '. . . it is fundamentally necessary to unite all available forces for conduct of the main operation in the Southern sector, with the aim of destroying the enemy west of the Don, so as subsequently to capture the oil regions in the Caucasus and cross the Caucasus range.'[4]

He went on to say: 'in any event, the attempt must be made to reach Stalingrad itself, or at least to remove it from the list of industrial and communications centres by subjecting it to the action of our heavy weapons.' A glance at the map shows that, although it was vital for the Germans to defend their northern flank by plugging the gap between the Don and Volga rivers, it was not actually necessary to capture Stalingrad itself which lay south-west of the Volga and was accessible to the Soviets only by river boats which would have been under constant artillery fire.

It is not necessary in this consideration of urban combat to look in detail at the strategic picture in

Soviet machine gunners engage the German invaders
amid the ruins of Stalingrad

German troops advance at Stalingrad

Guardsmen of the Red Army firing at a sector of
Stalingrad held by German soldiers. Step by step,
sectors of the city were cleared of German forces
using heavy automatic weapons such as this

this theatre of war, or at the decisions which ultimately led to the failure of Hitler's offensive in the
Caucasus. Suffice it to say that for a combination
of reasons Stalingrad began to assume a role centre stage. For one, the very name 'Stalingrad'
meant that, in Stalin's eyes, the city could not be
lightly given up. Moreover the city was an important industrial centre stretching for 25 miles along
the west bank·of the Volga, sustaining a population of 600,000 and providing work in steel, armaments and tractor factories. Stalingrad, quite apart
from its symbolic importance, was a showpiece of
the Soviet Union. Stalin decided it would be
defended.

The German army group B, commanded by
Colonel General Freiherr von Weichs and consisting of a total force equivalent to thirty divisions, was ordered to attack Stalingrad on 23 July
1942. The Soviet forces in the Stalingrad area
comprised the 62nd and 64th Armies supported
by the 1st Tank Army. On 4 August Stalin
appointed a new commander on the Stalingrad
front. His name was Andrey Ivanovich Yeremenko, his rank colonel general, and his age only
thirty-nine. He was ambitious and he thrived on a
challenge. Yeremenko arrived in Stalingrad in the
morning of 4 August. He got there just in time to
be on the receiving end of the first serious German attack on the city, launched at 0430 hours on
the 23 July. By the morning of the 24th the city
was in ruins and thousands of its citizens lay dead.
The Luftwaffe had managed to mount the equivalent of two thousand sorties against Stalingrad.
Although the blocking of roads by fallen buildings
hampered the movement of Yeremenko's forces
to the threatened sectors of the front line, later
experience—for example, of the Western Allies
at Cassino and Caen—showed that the destruction of large buildings can assist a determined
defender by slowing down the attacker's progress
towards the centre of a defended city.

Still Stalingrad held on. On 2 September Stalin sent a personal message to Zhukov, the commander-in-chief in the area. It read in part:

'(I) require the commanders of the forces deployed north and north-west of Stalingrad to strike at the enemy at once, and go to help the Stalingraders. No procrastination is permitted. Procrastination now equals crime. Throw all aviation in to help Stalingrad.'[5]

On 5 September the 24th and 66th Armies attacked in yet another effort to pinch out the German salient between the Don and the Volga. The attack did not succeed, but the Germans had to divert some of their forces northwards to beat it off, and this took some of the pressure off the 62nd and 64th Armies desperately trying to organise some kind of defence line around the perimeter of Stalingrad.

On 12 September Lieutenant General Vasily Ivanovich Chuykov took over command of the 62nd Army defending Stalingrad itself. Yeremenko had dismissed his predecessor for beginning to withdraw units from the city without orders. Chuykov, who had previously been deputy commander of the 64th Army on the Stalingrad front, had been studying German battlefield tactics closely during his few weeks in action, and had decided that the Germans were competent soldiers but by no means invincible. In particular he believed they derived their extraordinary success from the excellent coordination of all the elements of a modern mechanised army: aircraft, tanks, artillery and infantry. He also believed that they disliked the realities of close combat and often avoided this by the application of firepower. He had therefore decided that the most effective way to fight the Germans was to keep as close to them as possible. In that way the German infantry would be forced to fight at close quarters, whether they liked it or not, and their high command would not be able to coordinate artillery and air support so easily because of the risk of engaging their own troops.

September saw continuous and bitter close-quarter fighting as the city was contested, building by building. Armed police, firemen and factory workers were installed in suitable buildings—fifty

This photograph shows part of the bitter street fighting which took place in Stalingrad in the autumn and winter of 1942–3, after Hitler's armies had swept over Southern Russia to this centre of heavy industry on the Volga. Here, Soviet troops advance through a factory building wrecked by artillery fire

to a hundred men in each—and given orders to defend them to the bitter end. There were many localised battles. Typical of many was that fought by a battalion of Rodimtsev's Guards at Stalingrad railway station. On the morning of 17 September they came under attack by a force of German infantry supported by about twenty tanks which drove them out of the station and surrounding buildings. They counter-attacked and retook the station, only to be driven out again. During the day the area changed hands four times, but by nightfall it was again held by the Guards. Scattered around them lay the hulks of burnt-out tanks and the bodies of hundreds of dead of both sides.

While the battle raged in Stalingrad, Stalin and his general staff began to see the opportunity to attack the long-exposed German northern flank along the Don. Crucial to the plan, however, was Stalingrad: German attention had to be kept concentrated on the city and therefore it must be

held. The Soviet operation would involve a giant pincer movement, and the consequent encirclement and destruction of the German VI Army, IV Panzer Army and as many Romanian formations as possible. Success depended on the 62nd and 64th Armies continuing to pin down a large German force in the Stalingrad area, but with the bridgehead on the Volga now consisting of only the northern part of the city this was becoming increasingly difficult. The bridgehead was indeed so small that almost all of it could be brought under fire from small arms. The gains or losses in battle were measured in yards, and the basic unit of combat was the single sniper, or the six to eight man infantry assault group armed with automatic weapons, hand grenades, Molotov cocktails and anti-tank weapons. Most of the fighting took place actually inside the buildings—to appear outside was instant death. As soon as an assault group had captured a building it was followed up by a reinforcement group which was armed with heavy machine guns and automatic weapons, mortars, anti-tank guns, crowbars, picks and explosives.

A picture radioed from Stalingrad, showing Soviet infantry advancing in the south-west sector of the city

Red Army defenders of Stalingrad armed with automatic rifles fighting invading troops, who had a foothold in the neighbouring houses. They continued to defend the town house by house, street by street, while the relieving army to the north-west battled nearer.

German infantry and armour engaged in street fighting in Russia in 1942

These tactics proved very successful for the Soviets; the Germans were not so quick to adapt to the special situation in Stalingrad.

The supreme crisis for the 62nd Army came on 14 October. The Luftwaffe flew nearly 3,000 sorties, whilst on the ground the XIV and XXIV Panzer, 60th motorised, 100th Infantry and 389th Infantry Divisions assaulted the tractor and 'barricades' factories in northern Stalingrad. Some German units reached the Volga and split the 62nd Army in places. Close-quarter fighting continued until 30 October, but the German attacks grew weaker. Incredibly the defenders of Stalingrad held out.

Meanwhile vast Russian forces were assembling to execute one of the boldest counter-strokes of World War II. Altogether the Soviet force comprised over one million men, with 13,541 guns and mortars, 894 tanks and 115 aircraft.[6] That the Germans were not sufficiently aware of this massive concentration of force is a tribute to Red Army security and deception. But von Paulus was able to mount one more attempt to capture Stalingrad before the trap was sprung. At 0630 hours on 11 November he attacked with seven divisions on a three-mile front. He was met head-on by the Russian defenders. By midday, though, the Germans had reached the Volga on a frontage of 600 yards in the area of the Red October factory, thereby cutting the 62nd Army into three. But by 18 November, elements of the 62nd Army were able to counter-attack, block by block, house by house, room by room. Then in the nick of time, on 19 November, Zhukov sprung his trap.

It would not be appropriate in a book on urban combat to document the progress of Zhukov's counter-stroke; suffice it to say that it was an operation of epic proportions that completed its encirclement of twenty German and two Romanian divisions—a total of 330,000 men—on 23 November. The important point is that it could not have been achieved without holding onto Stalingrad.

The tables were now turned; with the onset of winter the temperatures plunged—but the 62nd Army, from being a beleaguered outpost, became part of a ring of steel around the German VI Army in Stalingrad. During December about 80,000 men—a quarter of the encircled force—were lost

from wounds, hunger and sickness; but the Germans continued the battle. Vicious street fighting went on until 2 February when the last of von Paulus's troops surrendered.

Stalingrad is the classic example of urban warfare. It involved the most troops, it went on for the longest period, the casualties were the highest and the stakes were momentous. Stalingrad was one of the key events of World War II, and the Soviets would argue that it was the crucial turning-point. Certainly Hitler was dumbfounded by the surrender.

Training in Urban Combat in England 1940–2

When the troops returned from Dunkirk in 1940, the British Isles faced the threat of invasion from Hitler's forces across the Channel. It was a threat that was taken extremely seriously. The potential value of Britain's effort was quickly realised and, using the lessons of Calais, urban warfare schools were set up the length and breadth of the land using derelict or bombed areas of cities as training grounds. Techniques were at first primitive and equipment was rudimentary, but at least the problem was being addressed and large numbers of infantrymen were given a basic grounding in the techniques of urban combat. It is true to say that the sheer scale and intensity of street fighting that was subsequently to occur at Stalingrad and at Cassino was not fully appreciated—but then there was no reason why it should be. Nevertheless, the techniques that were learned in the schools during the period 1940–2, even though they were never put to use in England, were indispensable in Italy, France and Germany in the period 1943–5. Techniques evolved as problems were encountered. Pamphlets were written and a theory of urban combat emerged—hitherto it had not been regarded as a discrete form of warfare. The aim was to make every city, town and village in the south of England—and London in particular—a sponge which would soak up and devour invading German forces; all training was to that end. In fact such tactics suited the parlous state of the British Army. The Expeditionary Force had left most of its equipment in France, and there was a chronic lack of armoured vehicles, of motor transport, of artillery and of the other heavy equipment

Street fighting training in Sunderland in 1942

required to fight a mobile battle in open country-side. The Army retained, for the most part, its light weapons—its rifles, machine guns, mortars and PIAT hand-held anti-tank weapons. All these could be used in the urban environment.

As the threat of invasion receded in 1941 and to a greater extent in 1942, the Army High Command could be forgiven for wondering if it had all been a wasted effort. But when the world held its breath as the Stalingraders fought the Germans to a standstill in late 1942, they realised that all their training had not been in vain. When the tables turned in Europe, British soldiers would undoubtedly be facing the Wehrmacht in an urban environment.

A trainee demonstrates how to clear a room of the enemy, by throwing a grenade through a window whilst being held upside down by two colleagues standing on the roof. The 'trainees' were Army Physical Training Corps soldiers. Whether the average infantryman would have been up to this sort of approach is open to doubt

3 Urban Combat in World War II: 1943–5

Other than the British experience in Calais, the Western Allies had had little opportunity to practise the art of urban combat in the first three years of the war. The only British success on land had been in North Africa, as far removed from an urban environment as was possible. The Soviet achievement at Stalingrad was noted, however, and the lessons tucked away for possible future use when a foothold was gained on the continent of Europe. The invasion of Sicily and subsequently of the Italian mainland in 1943 provided the opportunity.

Sherman tanks prepare to open fire on German snipers at Ortona. On the extreme right of the picture an infantry officer is preparing to direct their fire

Ortona, December 1943

Ortona was typical of the many communities up and down the Adriatic which were originally established as coastal strongholds in medieval days when the maritime power of Venice dominated Mediterranean commerce. Huddled against the massive 15th-century castle which crowned a high promontory thrusting squarely into the sea, the Old Town with all its tall, narrow houses and dark, cramped streets merged into the more modern section which had grown up on the flat tableland to the south. This newer part of the town was laid out in a system of rectangular blocks, although only the main thoroughfares were wide enough to allow the passage of a tank. The buildings were packed wall to wall, and rose generally

to a height of four storeys. From the eastern edge of the town an almost precipitous cliff fell away to the small artificial harbour, which was enclosed by two stone breakwaters protruding far out into the water. A deep ravine west of Ortona restricted the townsite to an average width of 500 yards for about one third of its length from north to south. This natural impregnability on three sides against attack meant that the German defenders could concentrate on blocking the only possible approach—the route from the south (Highway Number 16) by which the Canadians were attempting to force an entry. On the outskirts of the town this road became the Corso Vittorio Emanuele, continuing northwards to the Piazza Municipale. From this central town square, overlooked by the great dome of the cathedral of San Tommaso, the Via Tripoli led out past the cemetery to emerge as the main coast road to the north.

It was the 2nd Canadian Brigade's unenviable task to take Ortona. The 2nd Brigade consisted of the 1st Battalion the Loyal Edmonton Regiment, the 1st Battalion Seaforths, and the Three Rivers Tank Regiment.

The Edmontons, attacking northwards on a two-company front, spent the whole of 21 December clearing the score or more of scattered buildings which spread across the southern outskirts of the town. By nightfall they had reached the Piazza Vittorio, at the beginning of the main built-up portion of Ortona, and where they were within a quarter of a mile of the central square. The Seaforths' C Company had a stiff fight to clear the church of Santa Maria de Constantinopoli in the extreme south-east corner of the town; and during the afternoon the brigade commander ordered the full battalion to be committed to assist the Edmontons in what was developing into a major task.

Edmonton patrols reporting before first light on 22 December disclosed the effectiveness of the defender's demolition plan. The Corso Vittorio Emanuele was free of barricades for 300 yards or more, but all other lines of advance were blocked by the debris of houses which German engineers had systematically toppled into the narrow streets. It looked as if the defenders intended to channel the Canadian attack along the main street to the open Piazza Municipale, which presumably they

hoped to make a 'killing ground'. Lieutenant Colonel J. C. Jefferson, commanding the Edmontons, decided to clear the enemy from both sides of this central route so that the street itself might be swept of mines to enable tanks to penetrate the town. A Company took the left, and D the right, with B Company carrying out flank protection in the troublesome area between the main Corso and the esplanade overlooking the harbour. Company tasks were divided into platoon and section objectives, and commanders instituted a strict system of reporting each house clear before starting on the next. House by house and block by block the infantry worked forward, followed by the armour. By nightfall the Edmontons had reached the Piazza Municipale, although 25 yards short of it a high pile of rubble had stopped further advance by the tanks.

The day, typical of those to follow, had been one of bitter struggle against a stubborn and well organised defence. The German paratroopers, fresh, well trained and well equipped, fought like disciplined demons. Each sturdy Italian house that they elected to defend became a strongpoint, from every floor of which they opposed the Canadian advance with fire from a variety of weapons. They left other buildings booby-trapped or planted with delayed charges; and if these faced houses which they were holding, they demolished the front walls in order to expose the interiors to their own fire from across the street. Every obstructing pile of rubble was covered by machine-guns sited in a second storey, and the litter of shattered stone and broken brick usually concealed a liberal sowing of anti-tank and anti-personnel mines.

Although the employment of armour in such restricted conditions was anything but orthodox, the Three Rivers tanks gave the infantry invaluable support. They became in turn assault guns, their 75-millimetre shells smashing gaping holes in the walls of enemy-held buildings, and individual pillboxes, covering a sudden sally by the Canadian infantry with sustained bursts of machine-gun fire; frequently they carried ammunition forward and evacuated casualties through the bullet-swept streets. They performed these tasks under constant threat from German anti-tank guns sited to cover the obvious approaches

A Sherman tank firing its main armament at a target to its front

and often concealed behind the barricades so as to catch the attacking tank's exposed underside as it climbed over the rubble.

From the Piazza Municipale the Edmonton commander continued towards the Via Tripoli, intending to cut off the garrison holding the north-eastern portion of the town. Early on the 23rd, a troop of tanks managed to scale the barrier blocking the Corso Vittorio Emanuele; yet even with this support it took the infantry the whole of that day to cover the 200 yards to the wide Piazza San Tommaso in front of the cathedral. On the right, another Edmonton company had by night-fall secured the south end of the Corso Umberto I, which, being free of buildings on its seaward flank, could be covered by Canadian anti-tank fire and on that account offered a promising means of approach to the castle.

Their heavy casualties had left the Edmontons in a weakened state; none of the infantry battalions of the 1st Division had received any reinforcements since crossing the Sangro, and Jefferson was reduced to fighting on a basis of three companies of sixty men each. On 22 December, D Company of the Seaforths had been given the task of clearing the left flank, and when the remainder of the battalion entered the battle on the 23rd, this further eased the heavy strain on the Edmontons. The two commanders divided the town between them—the Seaforths to take the western half, and the Edmontons to push along the Corso Umberto I to the castle and the cemetery beyond.

In their grim efforts to advance, the infantry received magnificent support from the anti-tank guns—which were more suitable than field guns for shooting at such short ranges. The battalion 6-pounders and the 6- and 17-pounders of the 90th Anti-Tank Battery were employed with devastat-

ing effect against German occupied buildings. In the early stages of the fighting, two 6-pounders covered the advance of the infantry and tanks along the Corso Vittorio Emanuele, each firing high-explosive shells into the windows along either side of the street immediately ahead of the leading troops. When enemy fire prevented the sappers from demolishing the barricades of rubble which blocked the streets, the same guns blew the crests off the piles, enabling the Canadian tanks to mount them. As resistance in the streets stiffened, and infantrymen had to battle their way forward house by house, anti-tank gunners were called on to smash a way through. They obtained particularly effective results by first penetrating an obstructing wall with an armour-piercing round fired at close range, and then sending a high-explosive shell through the breach to burst inside. German snipers posted on the tops of buildings received short shrift: one round of a 6-pounder

Canadian infantry pick their way through rubble-strewn streets in Ortona

was usually sufficient to blast a tile roof to pieces. From a ridge south-east of the town two 17-pounders, firing at a range of 1,500 yards, systematically ripped apart buildings which the infantry indicated along the sea front.

Major E. J. Bailley, a Canadian Padre, conducts a short service at the grave of a German sniper on 22 December 1943 at Ortona

Failing in their attempt to outflank the enemy by striking up the Corso Umberto I, the Edmontons reverted to their former practice of working forward house by house. Having to get from a captured house to the next one forward, without becoming exposed to enemy fire along the open street, produced an improved method of 'mouse-holing'—the technique of breaching a dividing wall with pick or crowbar, and taught in battle-

drill schools from 1942 on. Unit pioneers would set a 'Beehive' demolition charge in position against the intervening wall on the top floor, and explode it while the attacking section sheltered at ground level. Before the smoke and dust had subsided the infantry would be up the stairs and through the gap to oust the enemy from the adjoining building. In this manner the Canadians cleared whole rows of houses without once appearing in the street; and as they progressed the German paratroopers automatically vacated the buildings on the opposite side.

The German Tenth Army war diary on the 23rd disclosed an attack by 'two battalions, supported by flame-throwers and 17 tanks . . . used as artillery', and reported that 'the number of our own casualties has compelled the abandonment of the more remote and southernmost positions . . . after exceedingly hard fighting'. Next day the 1st Parachute Division reported that 'in hard house-to-house fighting the enemy advanced to the centre of Ortona'. In fact the only troops to use flame-throwers in Ortona were the Germans. It took eight days in all to clear Ortona. The German tactics were an object lesson in defensive urban combat, and it is worth looking in some detail at how they set about defending Ortona, and at how the Canadians overcame the determined defence of that small town.

The Germans did not have to construct pillboxes—anyway there was probably not the time—because the sturdy Italian town houses afforded natural strongpoints. Buildings were methodically blown across streets to form barriers to provide covered approaches to exposed positions. Any buildings which overlooked a German position or which might have offered cover for the attacker were destroyed. In the buildings opposite German positions the front walls were demolished, thus exposing their interiors to fire from across the street. All roads, except those leading into the pre-selected 'killing grounds', were blocked by demolished houses. These piles of rubble were usually in such a position that they could be covered from above as well as from the rear. They were also usually liberally sown with mines and booby traps, easy to conceal amongst the dust and bricks. Houses which were not occupied were booby-trapped or had delayed

charges placed in them with timed fuses. In one instance a complete platoon of Canadian troops became casualties in this way, shortly after occupying a house. Retribution was swift. The Canadians wired up a small house with a large amount of explosive and carried out a limited withdrawal. The Germans reoccupied the building, but became victims of their own ruse: the resulting explosion accounted for some twenty German paratroopers. Both sides had learned a lesson. Thereafter the Canadians were forced to employ a larger number of troops than was strictly necessary for each attack. And each building had to remain occupied once it was captured, and held until the surrounding area had been cleared.

The main tank approaches were covered by German anti-tank guns sited at short range. Some were placed close up to the barricades to catch a tank's exposed underside as it lumbered over the rubble. Other guns were kept mobile and were rapidly shifted from street to street to counter a particular armoured threat. One very successful German tactic was to catch a tank in enfilade as it passed one of the openings to the innumerable narrow alleys which criss-crossed the main streets. A lookout would signal that a tank was about to appear, and a projectile was on its way the instant the tank showed its nose around the corner. Providing the tank was not travelling too fast—unlikely in the rubble-strewn streets—the chances were that the round would reach its target before it disappeared around the opposite corner.

In the face of such a determined and well organised resistance, the Canadians had to plan their tactics carefully. When the Edmontons reached the outskirts of the town, where street fighting started, they were advancing on a two company front approximating to the width of the town. The main road acted as the inter-company boundary. As the fighting developed the brigade commander realised that, in order to keep tight control and to maintain the fighting efficiency of the brigade, he would have to shorten his front and limit his objectives. So he committed the Seaforths to help clear the town and the main road was instead designated the inter-battalion boundary dividing the town in half. Neither battalion had a front of more than 250 yards.

Each battalion divided its respective sectors

A Canadian tank commander, wounded by a German sniper, is led to safety by a medical orderly.

Suddenly he collapses, and is given urgent medical attention

into various sub-sectors—which became company objectives—and company commanders again divided these into a series of platoon objectives. These might consist of not more than two or three houses. Rigid control was essential, so much so that when a company commander had completed the occupation of an objective he was not permitted to make any further advance without first reporting back to Battalion HQ. This was strictly adhered to, despite the strength of the opposition in any one sector. Platoon commanders never committed more than a section at a time. When given a street to clear, they placed their fire groups so that they covered both sides of the street, usually from the top floors of the first houses either side of the start of a street. Manoeuvre groups would then work their way from house to house rather than along the street. In some cases 'mouse-holing' was necessary—progressing from house to house by knocking holes through outside and intervening walls. But this was a painfully slow process, and could often be avoided by using courtyards and balconies to get from one house to another.

Every possible support weapon was used whenever and wherever possible. For example, the Edmontons were supported by 17-pounder anti-tank guns sited on a ridge south-east of the town which, at a range of between 1,000 and 2,000 yards, literally ripped the buildings apart in front of the advancing Canadians. Tanks were invaluable in spite of the lack of space for manoeuvre. They were used by the Canadians singly and in small numbers as assault guns or as static pillboxes. They were also invaluable for transporting ammunition and mortars forward to the fighting troops, and for evacuating the wounded over the bullet-swept ground. Several were lost due to German anti-tank gun fire, mines and grenades, but this price was considered negligible compared to the value of the assistance they provided.

Although artillery was available, it was found to be relatively ineffective in this type of fighting. The close proximity of the combatants often reduced no man's land to the width of a narrow street. Such situations required accurate pinpoint shooting, which can only be provided by tanks and anti-tank guns in the direct fire role. Also the old stone buildings of Ortona proved to

be surprisingly resilient to artillery fire. The upper storeys were battered but only the pounding of the Allied heavy and medium guns had any effect on the lower floors. Therefore the main artillery tasks were the continual harassment of the coast road to the north of the town, and the destruction of strongpoints with the 200lb shells of the Heavy Regiment.

The advance in Ortona was slow, methodical and relentless. Although casualties on both sides were heavy, the German paratroopers were eventually shifted from the town. Ortona was won by dogged house-to-house fighting—urban combat in its purest sense.

Cassino

Cassino is perhaps notable in the annals of urban warfare because so many different nationalities fought there. There were units from the US Army, from Britain, New Zealand, India, Nepal, from the French North African colonies of Algeria and Morocco, from Poland and of course from Germany. Cassino was part of the Gustav Line which stretched across Italy south-east of Rome and which had been chosen in January 1944 by Field Marshal Albert Kesselring, the Commander-in-Chief South, as the line on which the German Army would make its stand. An attempt to outflank the Gustav Line by landing two US divisions at Anzio on the coast south of Rome on 22 January 1944 was successful in engaging German reserves, but it failed to break out from the bridgehead. It therefore became important to link up with the Anzio force by driving up Route 6 along the Livi valley. To achieve this, the Gustav Line would have to be penetrated at the mouth of the Livi valley guarded by the town of Cassino and the Benedictine monastery on Monte Cassino above the town.

The first serious attempt to take Cassino got under way on 15 February. The responsibility was given to the New Zealand Corps commanded by General Bernard Freyberg. The plan was to seize the railway station at the southern end of the town, whilst the 4th Indian Division would capture the monastery above the town. The attack was preceded by a massive bomber raid, and the battle raged on the outskirts of the town for three days. The New Zealanders actually reached the

The monastery above Cassino after it had been subjected to intensive allied air and ground attack

station, but they had been unable to get their anti-tank guns across the river and when their PIAT ammunition ran out, their tenuous hold on the town was dislodged by a ferocious German counter-attack. At 1600 hours on 18 February the Maori survivors staggered back across the river; of the two hundred men who had started out, only seventy returned. On 22 February General Von Senger und Eterlin, commanding the XIV Panzer Corps, awarded General Baade, the German sector commander at Cassino, the Oak Leaves to his Knight's Cross.

The next attempt to take Cassino was codenamed Operation DICKENS. It began on 15 March with a strike by five hundred aircraft dropping 1,000 tons of bombs on an area measuring 1,400 yards by 400. As the last bombers droned away, Cassino was entirely laid waste. Not

(Opposite, above) THE BOMBING OF CASSINO   This picture was taken on 15 March 1944, after 2,500 tons of bombs had rained down on Cassino in Italy. As the barrage died down, the fight for the ruins began. New Zealand troops, leading the way, knocked out the enemy strongpoint in the ruined Hotel Continental in the south-west corner of the town, then stormed and captured Castle Hill (top right)

(opposite, below) New Zealand troops clambering over the ruins of bombed buildings, in order to close with the German defenders of Cassino

a single building remained intact—where the odd ruin did remain unsteadily erect, the artillery barrage which followed the bombers soon demolished it. The ensuing ground assault was again undertaken by the New Zealanders. Two companies of the 25th Battalion followed the artillery barrage into the town, with B Squadron of the 19th Armoured Regiment behind them. But the German paratroopers had survived the terrible bombardment and emerged from the ruins to fire on the New Zealanders whose radio communications failed at the critical moment. German snipers picked off the runners who carried messages, and the linesmen who were laying cable behind the advance. The tanks were unable to follow the infantry who were by now clambering over piles of concrete and girders and crossing 60ft craters. Smoke and dust made it difficult to keep direction, and progress was impeded even more by small parties of German paratroopers

German paratroopers fighting among the ruins of the monastery of Monte Cassino in April 1944

who kept continually on the move, firing from caves and ruined buildings on the hill above the town.

In particular the New Zealanders missed the close support of the tanks which could not get into the town because of the craters. By now it was appreciated that in street fighting tanks were decisive when they worked in close cooperation with infantry squads. The best team-work was at the lowest level when a pair of tanks covered each other while one of them fired several rounds into a building seconds before the infantry squad rushed in with grenades. But at Cassino, the tanks seldom got close enough to the infantry to locate either them or the enemy precisely, because of the rubble.

Intense fighting continued through the 16 and 17 March with gains of a few yards here and there. On 18 March the 28th Battalion was required to attack through the now bogged-down 25th Battalion—but they were attacking the Germans at their strongest point, and the German parachutists still held on doggedly. By now the

Germans had managed to introduce two Mark IV tanks into the town which gave the New Zealanders much trouble. Also, German snipers concealed in the slopes above the town were quick to punish anyone who loitered in the vicinity. On 19 March seventeen Honey tanks of the 7th Indian Infantry Brigade's reconnaissance squadron and 760th US Tank Battalion, and the Shermans of C Squadron 20th Armoured Regiment made a further attack, but this failed for lack of infantry support.

It is not proposed to document the rest of the Cassino story because although Cassino and Monte Cassino were not finally taken until May, there was really no more street fighting. The positions in Cassino were then held, and provided a firm flank for the operation to take the monastery which began in May and which mainly involved Polish, French and US troops. But this was mountain warfare rather than urban warfare.

The Allies' efforts at urban warfare in Cassino were not a success. The New Zealanders were not experienced street fighters, whereas the German 1st Parachute Division had had the advantage of recent training at Ortona. Tanks could not accompany the battalions into the town in the initial stages of the assault, communications broke down and consequently there was little or no direct artillery fire. The Germans, on the other hand, were determined and experienced street fighters. Above all, Cassino showed that massive aerial bombing can be an actual hinderance in an urban environment. The Allies still had many bitter lessons to learn from that late winter and early spring of 1944 in Central Italy.

North-West Europe 1944–5

In June 1944 the Allies landed in Normandy. The British, on the left flank of the invasion forces, were faced with the large industrial town of Caen. Once again the 'steamroller' approach was adopted by the British who employed not only

German paratroopers waiting in the yard of a town house in Cassino for orders to counterattack, 21 April 1944

An anti-tank gun of the 3rd British Division covers a
street in Caen on 9 July 1944 to guard against a
German counterattack

heavy bombers and field artillery to flatten Caen
but also used the big guns of Royal Navy capital
ships off the Normandy coast. The Germans put
up a stubborn defence, but overwhelming Allied
air superiority and the fact that the US armies on
the right flank of the Allied invasion forces were
breaking out northwards towards Paris made a
prolonged stand in Caen impossible. Nevertheless, the battle for Caen can still be described as
urban warfare, albeit at long range.

Paris fell without serious fighting and Le Clerc's
Free French Forces entered their capital city
amid the tears and cheering of their liberated
compatriots. It is interesting to consider that had

(Right) A 3in mortar of the Kings Own Scottish Borderers in action in a Caen street on 10 July 1944

the German commander of the Paris garrison not disobeyed Hitler's orders to fight to the last man, the glories of Paris might not have survived World War II.

By October 1944, the rapid Allied advance into Germany that had followed the break-out from the Normandy beaches had slowed to a crawl. Increasing German resistance, and various logistical and communications problems which plagued the Allies, combined to cause problems in the US sector. Lieutenant General Omar Bradley's 12th Army Group occupied a frontage, with the 1st and 3rd Armies along the Siegfried Line, and the 9th Army facing the river Roer. The 1st Army was situated near Aachen. Major General

(Below) British soldiers rescue a little girl from the shattered remains of Caen. Indescribable devastation awaited British and Canadian soldiers when they entered the city

Middleton's VII Corps occupied that army's southern sector; its 88-mile front extended from Losheim in Germany, north through eastern Belgium and Luxembourg to where the Our river crosses the Franco-German border. Corps headquarters was located in the small Belgian town of Bastogne, itself the hub of seven roads and a railway.

Alarmed by the worsening situation in the east, Hitler decided to take advantage of the stalled Allied advance and launch an offensive in the west. The aim of the operation would be to recapture the important port of Antwerp whilst encircling and destroying the 21st Army Group. Middleton's VII Corps was directly astride the advance of the Fifth Panzer Army.

Following a heavy artillery bombardment at 0500 hours on 16 December 1944, the Germans launched their offensive, gaining surprise and immediate local successes in all sectors. By 17 December the Germans were within eleven miles of Bastogne. The 101st Airborne Division, which was resting and refitting in France after operations in Holland, was alerted to move to Bastogne. Its commander, Major General Maxwell D. Taylor, was in the United States and his deputy was in England. Command therefore fell to Brigadier General Anthony C. McAuliffe, the division's artillery commander.

Even as the 101st and its attachments were moving into Bastogne during the night of 18 December, the Germans had reached a point just 3km from the town where they collided with elements of the 101st. By 21 December, Bastogne had been completely surrounded. That evening Generals Manteuffel and Luettwitz composed their now famous surrender demand: it was delivered to the 101st on 22 December, and received McAuliffe's even more famous reply, 'Nuts'. And despite the fact that they were outnumbered by at least three to one, the Americans held off repeated attacks until they were relieved on 27 December.[1]

Bastogne was not true urban warfare because for the most part, US forces were dug in on the edge of the town fighting to prevent the German tanks achieving a break-in. But much of the fighting was in the suburbs, and lodgements were achieved by German forces in the town, if only

temporarily. Bastogne is a good example of how light infantry, properly dug in and supported by artillery, tanks and anti-tank guns, can hold superior mechanised forces. The urban environment of Bastogne was the ideal place to make a stand. And the 101st Airborne Division made history, too.

The Allied advance through Germany in early 1945 was characterised by much town clearing. The following story of one British brigade is typical of many during the clearing of the Rhineland in February 1945: the 153rd Highland (H) Brigade was required to clear two large towns, Gennep and Goch, against determined and organised resistance. By this time, the Allies had learned the lessons of Calais, Ortona, Cassino, Caen and Bastogne. Before the attack on both towns the brigade studied large-scale maps, enlarged air photographs, and particularly air photographs taken from low-flying aircraft from an oblique angle—these were especially useful in identifying the key buildings upon which it was likely the enemy would base his defence. They also learned that it was important to be dressed and equipped properly for urban combat. Troops should fight lightly clad and without either back packs or picks and shovels which tend to catch in window frames and doorways. Other than that 'a rifle and bayonet, the Bren, a liberal supply of grenades, stout hearts and a very high standard of leadership are all that is required'.[2]

The Germans' defensive plan for Goch is not clear, and probably there never was one beyond a hasty last-ditch defence; this was to be repeated right across Germany. Most of the defenders were rushed in at the last moment, and fought in loose battle-groups drawn from a variety of units. The defenders totalled between 1,200 and 1,500—their mission was to hold Goch at all costs.

The German philosophy when defending a built-up area was to select and prepare a number of bastions throughout the town, each held in company strength. Buildings had to be defensible and dominate likely enemy approaches. The ideal contained a good cellar, which helped the defenders to survive the preliminary air and artillery bombardment, provided good fields of fire, and possessed escape routes covered from view and fire. Suitable buildings for artillery and mortar OPs were also considered to be important. The bastions were linked by frequent patrolling. Small teams armed with rifles, machine-guns and Panzerfausts sought to locate and destroy armour which threatened to outflank. Particularly favourite targets were the flame-throwing Crocodile Churchill VII tanks.

Morale and fighting spirit varied from one battle group to another. On the whole however, the will to fight was surprisingly high, given the circumstances. The Allies had air superiority, but the Germans had little armour and most of the remaining best men had been creamed off to the Waffen SS, with many units reduced to cadre size. Although artillery and mortars were superbly used, they were overmatched. But fanatical leadership by a few, a sense of duty in many, and a justifiable pride in their professionalism nevertheless enabled the Germans to fight on without any prospect of eventual victory—and this meant that the Allies could never underestimate them.

On the night of 10/11 February, 1945, troops of 153 (H) Brigade (5 Black Watch, 1 Gordons and 5/7 Gordons) crossed the river Niers and attacked Gennep from the north-west. This was a complete success; the enemy was taken by surprise as their defences faced the river Maas and the assault caught them in the back. Subsequently 51 (H) Division was reinforced by 32 Guards Brigade, who captured Hommersem and later Hassum only four miles to the west of Goch.

The 153 (H) Brigade plan for the attack on Goch was very simple. 5 Black Watch was to secure the north-western end of the town up to the main square by first light on 19 February. 1 Gordons and 5/7 Gordons would then pass through, the former swinging south to secure a main road junction, and the latter the railway line to the east. The 5 Black Watch plan was equally simple: D Company would go in first, followed by B Company, and then A and C Companies would exploit beyond that.

At 0100 hours on 19 February, D Company started the 153 (H) Brigade attack after a 15-minute artillery preparation. Earlier in the evening, as part of the 152 (H) Brigade effort, 2 Seaforth had secured the start line and placed a bridge across the anti-tank ditch. The company moved up in well dispersed file along the road bounded by fields, using the ditches whenever necessary, and

secured their objectives and a group of four houses to the south with ease. They were over the anti-tank ditch before the enemy realised it.

B Company followed through and consolidated on the road junction. A and C Companies were close behind and moved up both sides of the main street. Their patrols reported that the main square was empty. At this point—at about 0600 hours—D Company was ordered to take up a position level with the square in front of C Company. D Company had some difficulty in negotiating its way towards the square. Machine-gun fire made the main street almost impassable, but progress through back streets was easier. The cellars and surrounding gardens produced a haul of sleepy and disorientated Germans, many of whom were happy to surrender to humane captors.

Amongst the prisoners at the monastery was Colonel Matussek, the wounded commander of Goch and his HQ. Opposition had been moderate, a good number of the enemy had been caught coming up to man positions, and 166 prisoners were captured; it looked as if 153 (H) Brigade would have a relatively easy task in Goch. However, this was not to be. German reaction was thorough once the men had got over their surprise. Heavy shelling, mortar fire, sniping, accurate machine-gun fire and determined infantry made movement in the town difficult. When 5/7 Gordons arrived to pass through at first light to secure the railway, they had to fight to reach the square which was their start line. Either through lack of numbers or because of the darkness, 5 Black Watch had been unable to secure all parts of its objective. By nightfall 5/7 Gordons had only managed to advance 200 yards beyond it. Craters and debris prevented the tanks and Crocodiles from getting forward to help. This situation was to remain largely unchanged until the Germans withdrew in the early hours of 21 February; at which stage all objectives were finally captured.

The Gordon Highlanders and the Black Watch found that the greatest weight of artillery fire is required before zero hour; afterwards, however, it should lift from the infantry objectives to the far outskirts of town, as it is disconcerting to advancing troops to hear explosions close in front of them. It also drowns out the noise of enemy snipers and makes them more difficult to locate.

Moreover the effect of a 25-pounder on a house was limited. On the other hand 4.2in mortars were invaluable, as their bombs penetrated to the ground floor.

As to whether to bomb or not before an attack, the attackers of Goch found that heavy bombing had few advantages: air photographs are no longer accurate, craters and rubble preclude the use of tanks, and the danger area for heavy bombs makes it impossible to rush the objectives as the last bomb falls. Nor is it possible to clear a house from the top if it is so damaged that it is difficult to gain access to roofs. In the words of the commanding officer of the Gordons: 'from our experience in clearing a town that has not been bombed to one that has been heavily bombed, there is little doubt the infantryman would ask the airman to go elsewhere, particularly as he does not kill or even frighten the defenders the infantryman is going to meet'.[3]

The greatest lesson that 153 (H) Brigade learned was that urban warfare is slow and extremely tiring. Objectives must be limited and the enemy should never be bypassed. And they also learned to use a terrifying new weapon to great effect: the flame-thrower.[4] The brigade were able to put all the lessons they had learned at Goch and Gennep to good use the following month during hard fighting in the streets of Rees am Rhein.

The final major battle of World War II was an urban battle: the storming of Berlin, masterminded and commanded by Marshal Zhukov. The Russians estimated German strength around and in Berlin at up to 1 million men, 10,000 guns and 1,500 tanks, and 3,300 aircraft. However, this total was reached by adding up identified divisions and presuming them to be at full strength. In reality, German strength was in the order of 300,000 men, many either very old or young and untrained, with about 650 tanks and 900-1,000 aircraft, all very short of fuel. Against this, the Russians had massed 2.5 million men, 41,600 guns, 6,250 tanks and assault guns, and 8,300 aircraft, an overwhelming majority—but one that was required to ensure that Berlin was taken. The

Soviet infantry fighting house by house through the streets of Berlin in April 1945

Civilians, with hands aloft, being passed by a British tank entering the town of Uelzen on the 18 April 1945

general plan for the Berlin operation involved a staggered attack by three Fronts with converging axes. The aim was to cut up and isolate German forces and at the same time to bring the war to a rapid end by taking Berlin, the heart of Nazi Germany. The 1st Belorussian Front had the task of surrounding and taking the city, with its 2nd Echelon pushing on to meet the US and British forces on the Elbe.

The German defences were in three main belts, the strongest being the easterly belt along the Seelow Heights overlooking the Oder. Zhukov planned to commit his Front to an attack with six armies in the 1st Echelon and one in the 2nd, supported by several artillery divisions of the High

Command Reserve and a Reserve Army of Polish troops. The attack was spearheaded by 3 Shock Army, which consisted of 12 Guards Rifle Corps; 7 Rifle Corps; 79 Rifle Corps; 9 Tank Corps; 11 Mechanised Corps; 1 Mechanised Corps; and had in support 5th Rocket Launcher Division.

The army planned a two-Echelon attack since the defence was in depth. Two corps—the 79th and 12th—went in the First Echelon, the 7th in the Second. The breakthrough sector was to be 6km. The tank corps were held back to be committed as breakthrough or exploitation reserves. In the rear of the army, as in every Soviet army, was an NKVD division to provide 'backbone'.

In Berlin, meanwhile, life in the city for its 2½ million inhabitants went on as normal: shops traded, banks issued money, and sixteen cinemas continued to function. The 15 April was a moonless, cloudless night. At 3am local time, 41,600 guns, mortars and rocket launchers opened up in a barrage on the defences, one gun firing every 3½ yards of the breakthrough sectors. The drama of the scene was intensified by the light of 140 searchlights, and the noise of 500 night bombers striking the German rear. Over 1¼ million shells were fired during the first day. The troops advanced into the dust cloud created by the shelling, fighting their way slowly and cautiously forward. In contrast to earlier battles, they showed little élan, and it required a great deal of political motivation to encourage the Soviet soldiers to face death at this stage of the war.

It took 3 Shock Army three days to break through the first lines of defences, and until 23 April to reach the suburbs. They entered the city from the north and east on four axes, and the troops broke down into shock groups of mixed infantry, tanks and sappers with heavy artillery support. Even 203mm howitzers were dragged into direct fire positions. When the rubble prevented Katyushka multiple rocket launchers from reaching their direct fire positions, the rockets

A King's Own Scottish Borderers bren gunner firing down a road during the entry into Bremen in northern Germany on 25 April 1945

A Soviet anti-tank gun in action in the streets of Berlin. These weapons were often fired at almost point blank range at tanks and strongpoints

were dismounted, carried by soldiers to the upper floors of houses, and used as primitive bazookas. Most leading companies had engineer flame-throwers attached.

Although the bulk of the city's defenders were irregular troops and there was no 'citadel' defence established, they inflicted enormous casualties on the Russians and slowed down their rate of advance considerably. By this stage, the Russians outnumbered the defenders by 30:1, but the confines of the city prevented them from deploying this superiority effectively. First to penetrate to the centre of the city was the 79 Rifle Corps with the 150, 171 and 207 Rifle Divisions. By 28 April, leading elements had reached the river Spree. In a desperate battle on the night of 29 April, the 1st Battalion of 756 Rifle Regiment of 150 Rifle Division managed to seize a crossing over the Spree, and two regiments, a tank brigade and a flame-thrower battalion crossed to attack the Gestapo HQ on the Southbank; this finally fell on 30 April at 4am. The Russians were now only 400 yards from the Reichstag, but they were too exhausted to press on, and moreover were open to flanking fire from a strongpoint in the Opera House. Consequently, 89 heavy field guns and a battalion of Katyushkas were brought up, and at 4.30am on 30 April opened fire on the Reichstag to cover an

(Right) Bremen proved hard to capture. Here, on the 26 April, South Lancashire Infantry are clearing German snipers from the railway sidings in Bremen

(Below) The Hammer and Sickle being raised on the roof of the Reichstag in Berlin on 30 April 1945 by Red Army soldiers

A 3in mortar crew from the 1st Battalion the West Yorkshire Regiment in action during the Burma campaign in Meiktila

assault by the 2nd Echelon regiments of the 171 and 150 Divisions; 207 Division attacked the Opera simultaneously.

The first assault on the Reichstag was at dawn but was beaten back by some 1,300 SS defenders.

The second assault at 1130 hours also failed. Finally, at 1830 hours, with tank and artillery support, soldiers from the Shock battalions of the 674 and 380 regiments broke into the building, and under cover of dusk, further troops were able to infiltrate. There followed a desperate battle between the SS defenders and the best part of two Soviet regiments. At 2150 hours, a group of Soviet soldiers made it onto the roof of the building with the Soviet flag. The battle for the building raged throughout 1 May, and it was not until nightfall on 2 May that the garrison in the Reichstag surrendered. That same night General Wiedling surrendered the Berlin Garrison to the Soviets.

Of the 2,500,000 men involved in the Berlin operation, the total Soviet casualties for the operation were 102,000 dead and 195,000 wounded; 36,000 of the dead fell within the last few days of the battle for the city. The battle for the Reichstag building alone cost 3 Shock Army over 2,000 lives. German casualties are not known precisely, but are estimated to be at least twice the Soviet number.

Urban warfare was a phenomenon that pervaded every theatre throughout World War II. There was vicious street fighting in 1945, for example, during the Burma Campaign in Mandalay. The battle for Berlin, for the heart and soul of the Third Reich, however, reached levels of intensity that showed very clearly that a nation will fight hard, even fanatically, to preserve its capital. The last significant battle of World War II was an urban battle.

4 Urban Combat since 1945

If World War II saw sustained urban combat for the first time in history, the years since have done nothing to change the trend towards fighting in built-up areas. A succession of conflicts have occurred since 1945 that have taken place mainly or to some extent in cities. Some examples are Palestine 1945–9, Cyprus 1945–59, Suez 1956, Algeria 1954–62, Vietnam, Aden 1964–7, Northern Ireland 1969–1991, the Philippines and especially the Lebanon spasmodically throughout the 1980s, and Bejing, Bucharest and Panama City in 1989. There are, of course, many other examples but these are some of the most interesting. This chapter is inevitably selective and can only look at some of the many instances of urban combat since 1945.

Palestine

The movement to create an independent Jewish state in Palestine took on added momentum as soon as World War II ended. A terrorist campaign against the British began in a small way in late 1945 but escalated with the murder of several British soldiers in early 1946. After a particularly brutal series of murders by the Jewish terrorist organisation, the Irgun, the British authorities planned a large-scale operation involving all British military units in Palestine to search and screen suspect Jewish houses and their inhabitants—this even included members of the Jewish Agency. Operation Agatha, as it was called, met with considerable success: in all some 600 weapons were captured. But the Irgun, not to be outdone, then pulled off their most outrageous coup to date. On 22 July they planted large quantities of explosive under the King David Hotel, used for government and military offices. Despite a telephoned warning of an impending explosion and a gun battle that had resulted when the terrorists were surprised in the act of planting the explosive, most of the occupants of the offices were not alerted. In the ensu-

ing explosion over ninety people were killed.

A four-day cordon and search operation of Tel Aviv followed involving 17,000 British troops. More arms were discovered and hundreds of suspects detained, although once again the elusive Menachem Begin, the leader of the Irgun and subsequently Prime Minister of Israel, escaped detection. By this stage, however, even the moderate Jewish Agency had determined that the only choice was to work for the removal of British rule. There had been one last chance of compromise the previous April when the Anglo-American Committee of Inquiry had recommended that 100,000 Jewish refugees be transferred from Europe to Palestine. The British had announced that they would agree to the plan, but only if both Arabs and Jews surrendered all their weapons. There was, of course, no chance of either side complying. By July 1946, 27,000 British troops were involved in Palestine. Only one year after peace in Europe, a vast British Army was again in action—but this time engaged in an impossible task which none of them relished. Although there were occasional incidents of indiscipline, the behaviour of the troops was remarkably restrained under conditions of extreme provocation.

During 1946 forty-nine British soldiers were killed and many more wounded. But worse was to come in 1947. In February that year Ernest Bevin, unable to find a middle way between Palestinian and Jewish demands, washed his hands of the affair and referred the matter to the United Nations. The British left Palestine to its fate in 1948. It was a sorry tale of political intrigue and of terrorism for political ends. What made Palestine significant was that it was the first of many postwar campaigns to achieve national independence. The Jews quickly realised that they could cause most disruption and mayhem in urban areas, and regularly exploited this particular vulnerability of the British Army. Unable to resort to the use of air

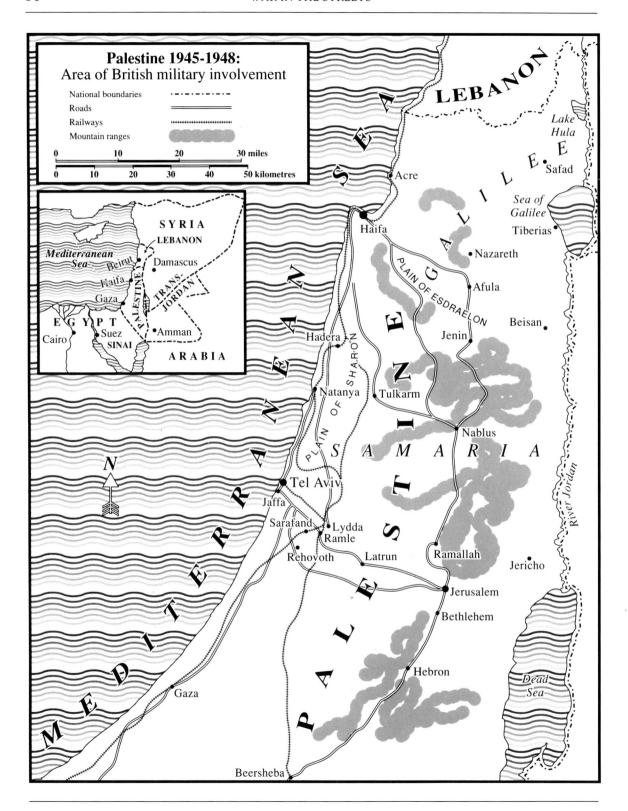

Palestine 1945-1948:
Area of British military involvement

National boundaries
Roads
Railways
Mountain ranges

0   10   20   30 miles
0   10   20   30   40   50 kilometres

MEDITERRANEAN SEA

LEBANON

Lake Hula

Acre

Haifa

GALILEE

Safad

Sea of Galilee

Tiberias

Nazareth

PLAIN OF ESDRAELON

Afula

Beisan

Jenin

Hadera

Natanya

PLAIN OF SHARON

Tulkarm

SAMARIA

Nablus

PALESTINE

Tel Aviv

Jaffa

Sarafand

Lydda
Ramle

Latrun

Ramallah

Jericho

River Jordan

Rehovoth

Jerusalem

Bethlehem

N

MEDITERRANEAN

Hebron

Dead Sea

Gaza

Beersheba

Mediterranean Sea

SYRIA

LEBANON

Beirut

Damascus

Haifa

Gaza

PALESTINE

TRANS-JORDAN

EGYPT

Cairo

Suez

SINAI

Amman

ARABIA

power or artillery, the British were immediately at a disadvantage in this urban environment.

Despite some experience of conventional urban warfare during World War II, the British Army had failed to translate its experience into the context of a terrorist campaign. However, this conflict marked the beginning of the post-1945 tendency to exploit the urban situation—and by coincidence, the next campaign involving urban combat and also involving the British Army took place only a few miles away across the sea, in Cyprus.

Cyprus 1954–9
The Cypriot campaign, initially for union with Greece—known as ENOSIS—and subsequently for independence from Britain, was for the most part fought in the mountains and forests of

The Jewish terrorist campaign

British troops training in Cyprus in 1989 in urban combat techniques. Despite the EOKA campaign for union with Greece, the island became independent and Britain retained residual military rights in the form of Sovereign Base Areas

Cyprus. It was here that EOKA (Ethniki Organosis Kyprion Agoniston, or Cyprus Freedom Organisation), had most of its support and it was here that General Grivas and his gangs hid out. Yet most of the headline-grabbing terrorist incidents took place in either Nicosia or Famagusta—publicity for a terrorist organisation is as vital for its continued existence as the air it breathes.

Archbishop Makarios became the focal point of the Greek Cypriot campaign to achieve ENOSIS. The British authorities therefore decided that one practical solution to the problem was to remove him from the scene altogether. On 9 March 1956 Makarios was arrested at Nicosia airport en route to Athens, and instead put on an aeroplane to the remote British territory of the Seychelles.

Makarios' deportation coincided with a dramatic intensification of hostilities by EOKA. During the next three months, two members of the security forces, on average, were killed each week. The army was forced onto the defensive, the safeguarding of life and property taking priority; inevitably offensive operations against guerillas were neglected so as to keep order in the towns. Curfew enforcement became part of the daily routine of a soldier's life. Then in the second half of March, five British soldiers were killed and an unsuccessful attempt was made on the life of the Governor General, Field Marshall Lord Harding, by a Greek Cypriot member of his staff who placed a bomb in his bed. And a further dimension was added to the already complex situation when serious inter-communal violence broke out, sparked off, amongst other things, by EOKA including Turkish policemen as legitimate targets. On 27 April, British security at Nicosia Airport

Cyprus: scene of operations against EOKA 1954–9 and by UN forces since 1964

was shown up for the second time in two months with the destruction of a Dakota by a bomb. On the other hand, a series of successful operations in the mountains led to the defeat of EOKA outside the two large cities on the island, Famagusta and the capital Nicosia.

But the collapse of the mountain gangs was not followed by a parallel decline in violence in the rest of the island. In June, an English schoolteacher and a police officer were shot dead, and Mr Justice Shaw was seriously wounded by gunmen; six soldiers and eight Greeks, the latter branded as traitors or collaborators, were also killed. On 16 June EOKA threw a bomb into a restaurant and (unintentionally) killed the American Vice-Consul. During July and August the violence continued on the same scale, with twenty-nine more soldiers, officials and Cypriots dying, all these attacks taking place in urban areas.

The British government now decided to try a more conciliatory tone. As part of the new approach it was decided to retire Harding and replace him with a colonial servant of more liberal tendencies, Sir Hugh Foot. Harding departed on 3 November. But by the time Foot arrived four

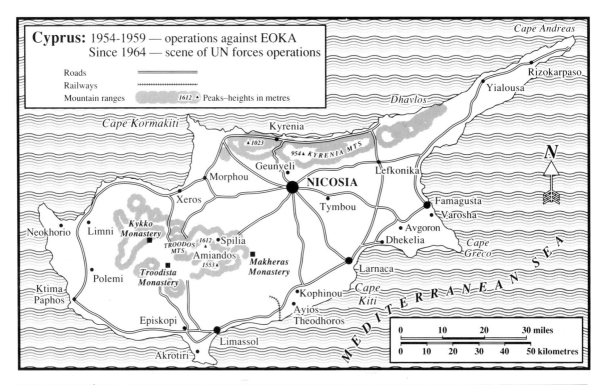

Cyprus: 1954-1959 — operations against EOKA
Since 1964 — scene of UN forces operations

Roads
Railways
Mountain ranges   1612 ▲ Peaks–heights in metres

weeks later, EOKA had blown up a Canberra bomber at Akrotiri Airfield, an ominous foretaste of what was to come. Grivas mistrusted Foot's motives; and the Turkish community demonstrated its disapproval of the new governor's plans for self-government and eventual self-determination by rioting in Nicosia on 27 January 1958. The incident precipitated a pitched battle between British soldiers and thousands of Turkish youths, in which seven Turks were killed and twelve soldiers injured.

In late March the EOKA campaign was resumed. Fifty bombs exploded during the first ten days of April. Then on 16 April, Foot asked a Greek contact to arrange for the delivery of a personal letter from himself to Grivas in which he appealed to the EOKA leader to suspend his campaign of violence and sabotage, to save the people of Cyprus further agony. He even offered to meet Grivas alone and unarmed. But Grivas prevaricated; he ordered a stop to the violence but did not reply to Foot's letter, suspecting the authorities wanted an answer as an aid to tracing him in his hideout.

On 4 May two British soldiers were shot dead as a reprisal for alleged ill-treatment of detainees held in the controversial prison camp, Camp K. Meanwhile the Turkish community, whipped into a frenzy by broadcasts from Turkey calling for the partition of Cyprus, campaigned vociferously for partition. Violence between the two communities broke out in early June and reached a climax when eight Greeks were massacred in a cornfield near the Turkish village of Geunyeli. The burden of keeping the two communities apart had fallen to the British troops; and when on 17 July EOKA murdered five Turks, Foot, in desperation, ordered them to make an island-wide swoop on 21 July in a last-minute attempt to avert civil war. Fifty Turks and more than 1,500 Greeks were arrested. In under two months, ninety-five civilians had been killed in the inter-communal violence and many more injured, the vast majority of these being in Nicosia and Famagusta.

By now EOKA had realised that a terrorist campaign in the cities would be more effective on the world stage than a guerilla war in the mountains. The assault against the army reached its worst in the summer and autumn of 1958. In August, Sergeant Hammond was shot dead in Ledra Street in Nicosia (or 'Murder Mile' as the press had christened it) while walking with his two-year-old son. In October Mrs Mary Cutliffe, married to a senior NCO in the Royal Artillery and mother of five children, was shot dead while shopping in Varosha; Mrs Robinson, another sergeant's wife, was seriously wounded. This was too much. In less than two hours, over a hundred Cypriots had been rounded up and taken to detention centres for questioning.

There is little doubt that the troops were so disgusted by this horrific murder that they broke some of the rules. One man, aged thirty-seven and the father of six children, Panayotis Chrysostomos, was found dead at brigade headquarters. He had been arrested by the Royal Ulster Rifles in Famagusta, and the subsequent enquiry found that Chrysostomos had died of heart failure while suffering from respiratory complications arising from the fracture of seven ribs. However, because the evidence of so many of the witnesses conflicted, the coroner was unable to determine how the ribs had been broken. Another Cypriot, Andreas Loukas aged nineteen, died of severe head injuries at Karaolis Camp. Again evidence was conflicting, and the coroner found that Loukas had been murdered by persons unknown.

An estimated 256 people were injured in the round-up. At Loukas's inquest Mr Justice Trainor commented that 'People were so assaulted and beaten that doctors were fully occupied at Karaolis Camp and the hospital tending the wounded that evening. One can fully understand the horror, disgust and anger that filled the hearts of everyone on that day, but nothing can justify the assaults on persons who had done nothing to warrant them'. In the event Mrs Cutliffe's killer was never found. Although EOKA's cause was severely damaged by this appalling incident, attacks against British troops were maintained throughout the autumn and early winter. All the forces were targeted—on 11 November two RAF men were killed in a NAAFI establishment by a bomb placed by Greek Cypriot employees.

It was at this juncture that Cyprus was again discussed in the UN, and the degree of accord reached between the Greek, Turkish and British representatives led to further Greek–Turkish

meetings. These culminated in February 1959 with a meeting in Zurich between the Greek and Turkish prime ministers, who finally announced that they had reached agreement on a general plan for a settlement. Independence was now seen by all parties as the only realistic solution. ENOSIS and partition were dead, and the Zurich plan provided for a Greek president and a Turkish vice-president.

Legislative authority was to be vested in a House of Representatives in a ratio of 70 per cent to 30 per cent Turkish members. The judiciary, civil service, local government and the military were similarly divided. The Zurich agreement was ratified by the British Government on 19 February 1959 in London. On 1 March Archbishop Makarios was permitted to return to Cyprus. On 9 March Grivas, who had not liked much of the agreement, reluctantly ordered a ceasefire; on the 17th he was flown unobstrusively to Athens where he was given a hero's welcome.

As a guerilla leader Grivas was in some respects remarkably successful: he tied down 40,000 British troops and killed ninety-nine; he avoided capture in a small island for years on end. But in a wider sense he achieved very little; he perhaps brought colonial rule to an end a few years earlier than would otherwise have been the case, though other factors had made, or were making, Cyprus strategically redundant with or without EOKA— Suez, changing treaty obligations, and longer-range and faster aircraft. Grivas's greatest achievement was to switch EOKA's main effort from a rural guerilla war to urban terrain, this against his own inclinations as a mountain guerilla leader. It was the urban threat that forced British hands.

Algeria 1954–62
At much the same time as the British were fighting EOKA in Cyprus, the French were involved in a bitter urban conflict in Algeria. The Algerian war

Today, British troops train in the art of urban warfare in a training ground in Cyprus. Their rifles are fitted with a device which fires a laser beam at its target. Detectors can be seen on the soldier's helmet in the foreground: if a player is 'hit' an audible alarm is sounding by the helmet-mounted equipment. This results in extremely realistic training

lasted nearly eight years, longer than both World Wars I and II. Compared to many of the British colonial wars, its results were arguably historically more important and its potential for real disaster greater. It was to cause the fall of six French prime ministers and the collapse of the Fourth Republic; it came close to bringing down General de Gaulle and the Fifth Republic and, most significant of all, it very nearly plunged metropolitan France into civil war. It was an appallingly savage conflict which resulted in the killing of probably a million Algerians and the expulsion from their houses of approximately the same number of European settlers. Although the Algerians practised unspeakable mutilations, the French settlers—the so-called 'Pieds Noirs'—also tortured their victims mercilessly.

It is not possible to explore here the full complexities of the Algerian conflict. The war was immensely complicated. It was, to some extent, a classic war of independence fought by the native Algerians against the French colonial power, but it was also a war between the Pieds Noirs who wished to *keep* Algeria French, and successive French governments which were prepared to grant independence at some stage. Without going into the minutae of the war, it is worth noting that virtually all the headline-grabbing episodes of this most unpleasant and bloody war took place in Algiers, a vast and sprawling Arab city. The 'Front de Libération Nationale' (or FLN) had declared its determination to achieve national independence on All Saints Day 1954. By 1956 the movement had appreciated the importance of cities in its liberation struggle. The Arabs knew every inch of the tortuous alleys of the Casbah, so narrow that it is possible to jump from one roof top to another, and where one square kilometre housed a teeming population of 100,000 Muslims. FLN operatives quickly realised the potential for turning the Casbah in Algiers into a veritable fortress from which a campaign could be launched: with the aid of skilful masons, they organised a whole series of secret passages leading from one house to another, created bomb factories, arms caches and virtually undiscoverable hiding-places concealed behind false walls.

The FLN bombing and assassination campaign of 1956 aimed to kill as many French settlers as possible; bombings were indiscriminate and caused appalling carnage in the heavily populated city. Then on 28 December Amédée Froger, the mayor of Algiers, was assassinated by the FLN. The next day the whole of Pieds Noirs Algiers turned out for the funeral procession of the murdered leader. Tension reached breaking point when a bomb was detonated inside the cemetery—had the cortège arrived on time, it would have killed many innocent mourners. The crowd went berserk. Innocent Muslims were dragged out of their cars and lynched; veiled women were smashed on the head with iron bars—four were killed and fifty injured. As a result of this outrage, the Parachute Regiment were granted full responsibility for the maintenance of law and order in the city, and by the autumn of 1957 had ended the grim battle of Algiers in undoubted victory. But it was a hollow victory. The excesses of the Paras, and the killing and torturing of Muslims, had started a drift away from France of the middle ground—the liberals and the thinking middle classes. Furthermore the resettlement of over a million Algerian peasants into barbed-wire encampments looking much like concentration camps caused many more Frenchmen to be alienated.

The final years of the Algerian war were dominated not by the Muslims or the FLN but by the Pieds Noirs themselves. In September 1959 de Gaulle offered Algeria self-determination. This led to the formation of an organisation known as the Front National Français, or FNF, which determined to keep Algeria French. In January 1960 Frenchmen actually fired on Frenchmen during what became known as 'Barricades week' in Algiers. In a confrontation between FNF supporters and gendarmes, 6 demonstrators and 14 gendarmes were killed, and 24 and 123 wounded. In the next two years France came close to civil war: the Organisation Armée Secrète or OAS was formed, and even attempted to assassinate de Gaulle in its efforts to keep Algeria French. The rest is history: de Gaulle triumphed, civil war was averted, the OAS was defeated, and independence was granted to Algeria in September 1962.

The importance of this campaign is that the war was fought more than anywhere else in the streets of Algiers. This is where it was won and lost: this is where the vast majority died: 259 in June 1958, 184 in January 1959, and 143 in December 1959 for instance. This is where bomb attacks were carried out, kidnappings took place, and the barricades were erected. The cost in human life was considerable: according to French figures their forces lost 17,456 dead, 64,985 wounded and 1000 missing. European civil casualties are put at over 10,000, with 2,788 killed. French estimates put Muslim dead as 141,000 male combatants killed by the security forces, and a further 78,000 killed by the FLN. Virtually all casualties occurred in Algiers. It was, by any standards, urban combat on a grand scale.

It was perhaps only a matter of time before Arab resentment boiled over elsewhere in the region. The next urban upheaval was in Aden.

Aden 1964-7

If a particular incident is needed to mark the beginning of the urban campaign in Aden, then it happened on the evening of 23 December 1964 when a grenade was tossed through the window of an RAF officer's quarter where a teenage party was taking place, killing the sixteen-year-old daughter of an air commodore and wounding several other children. The campaign to oust the British from Aden and destroy the embryo federal government was Egyptian-inspired and this, its second and much more serious phase, had started as it meant to go on. There had in fact been 36 terrorist incidents during 1964 in Aden, but this jumped to 286 in 1965, 510 in 1966 and nearly 3,000 in the ten months before the British departed in 1967.

The geography of Aden provided the perfect scenario for a terrorist urban war in miniature. Confined within the boundaries of a relatively small and well defined town, the world was able to watch with ill-concealed fascination the blow-by-blow events that went on within this small and transparent arena. The terrorist campaign had been given added impetus by the announcement in the British government Defence White Paper of July 1964 that South Arabia would be granted its independence not later than 1968. However, although political and economic independence was planned, military independence was not, as the intention at this juncture was to maintain a British military base in Aden.

Once the date for independence was set, the Egyptians and the various national movements they supported set about gaining power for themselves by 1968. There were three main nationalist groups: the South Arabian League (SAL), the National Liberation Front (NLF) and the Front for the Liberation of Occupied South Yemen (FLOSY). The most formidable of these from the military point of view was the NLF because it was they who, from the start, firmly espoused violence as the best and most effective means of achieving their aims. The NLF was solely responsible for all the acts of terrorism in Aden State until early 1966, and for most of them after that. Both SAL and FLOSY, however, believed in political action to achieve their aims, though FLOSY also resorted to violence, not only against the British but against the Federation and Aden State too. These, then, were the organisations which the British Army was to face, and they followed all the traditional techniques and tactics of intimidation, industrial action and propaganda that terrorists before them had employed against the British elsewhere in the Middle East, in Palestine and Cyprus.

The geography of Aden State needs further explanation in order to understand the complex events of 1965-7. In the north, nearest to the land frontier with the Yemen and astride the frontier with Lahej, lay the twin towns of Dar Saad and Sheikh Othman, both of which were used by the terrorists as 'mounting areas' for many of their operations in Crater. Sheikh Othman in particular was regarded by the security forces as perhaps the most troublesome Arab township in Aden State. To the west, and adjacent to Sheikh Othman, was Al Mansoura which contained the detention centre where those suspected of terrorist activities were detained under the Emergency Powers Act.

Further west, and en route to Little Aden, was the newly constructed federal capital of Al Ittihad which, along with the BP oil refinery and the 24th Infantry Brigade military cantonment in Little Aden, presented an enormous security problem. All these installations had to be constantly guarded.

Across the isthmus, on which lay Khormaksar Airfield and Radfan Camp, was the town of Aden, itself divided into four constituent parts: Steamer

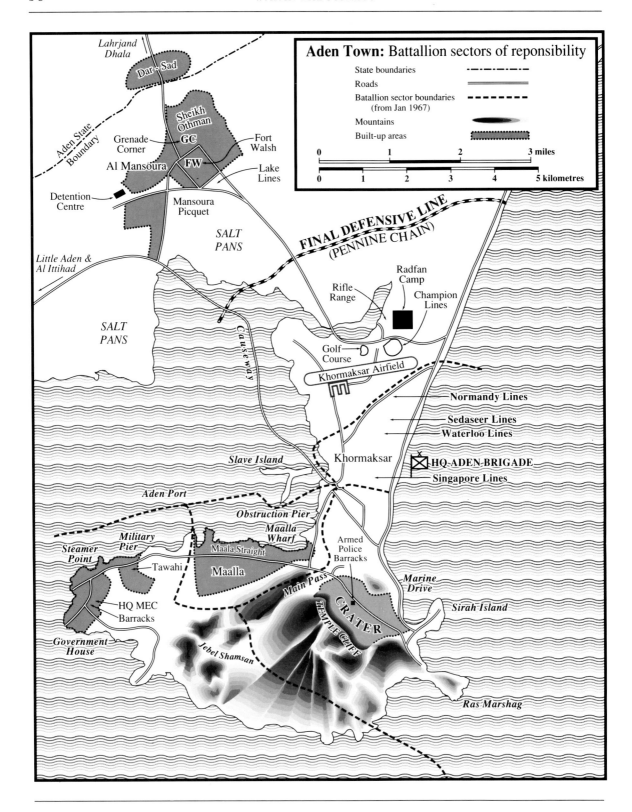

Aden Town: Battalion sectors of reponsibility

State boundaries

Roads

Batallion sector boundaries
(from Jan 1967)

Mountains

Built-up areas

0   1   2   3 miles

0   1   2   3   4   5 kilometres

Lahrjand
Dhala

Dar – Sad

Aden State
Boundary

Sheikh
Othman
GC

Grenade
Corner

Fort
Walsh

Al Mansoura   **FW**

Lake
Lines

Detention
Centre

Mansoura
Picquet

SALT
PANS

FINAL DEFENSIVE LINE
(PENNINE CHAIN)

Little Aden &
Al Ittihad

Radfan
Camp

Rifle
Range

Champion
Lines

SALT
PANS

Causeway

Golf
Course

Khormaksar Airfield

Normandy Lines

Sedaseer Lines
Waterloo Lines

Slave Island

Khormaksar

HQ ADEN BRIGADE

Singapore Lines

Aden Port

Obstruction Pier

Maalla
Wharf

Armed
Police
Barracks

Military
Pier

Steamer
Point

Maala Straight

Maalla

Marine
Drive

Tawahi

Main Pass

CRATER

Sirah Island

HQ MEC
Barracks

TEMPLE CLIFF

Government
House

Jebel Shamsan

Ras Marshag

Point, Tawahi, Maalla and Crater. Crater was predominantly an Arab town and was regarded by Arabs as the true capital of Aden State. Steamer Point contained Government House and Headquarters Middle East Command (HQMEC); Tawahi was the main business and shopping centre; and Maalla, as well as being an Arab township, contained many service married quarters.

In January 1965 Sir Richard Turnbull succeeded Sir Kennedy Trevaskis as High Commissioner to the Federation of South Arabia. He was in overall charge of defence and internal security matters, advised as necessary by the Commander-in-Chief Middle East, who delegated effective control of internal security to the GOC Land

Aden town: showing the battalion sectors of responsibility

The inverted arrow marks the spot of a Parachute Regiment Observation Post in the centre of Crater in Aden town, from which British soldiers kept watch on the city

Forces, who had at his disposal the Aden Brigade, the HQ of which was at Singapore Lines on the isthmus, and 24 Brigade with its HQ in Little Aden.

In early March one Abdul Mackawee, previously leader of the Opposition People's Socialist Party, was appointed chief minister of the Aden State government. An ardent nationalist, Mackawee ensured that every security measure promulgated by the High Commissioner was opposed and criticised, and that every accusation against British troops was upheld and encouraged. Finally on 22 April, he dissociated himself and his government from the curfew imposed by the High Commissioner following an attack on the Aden Commissioner of Prisons. By now it was apparent that more troops would be needed, and in late April the total garrison was raised to five battalions.

On 5 June the new GOC Land Forces, Major-General John Willoughby, was also appointed security commander, a new post designed to encourage greater co-operation not only between the two brigades in Aden but also between government, the police, the intelligence community and the other service chiefs. The following day the High Commissioner introduced emergency regulations in Aden State which enabled him to detain suspects without trial for up to six months, and to proscribe all terrorist organisations which he believed were engaged in subversion or terrorism.

On 29 August Arthur Barrie, a senior British police officer, was shot dead in Crater. Then ten days later Sir Arthur Charles, the British Speaker of the Aden State Legislative Council, was ambushed and killed by a gunman—also in Crater.

On 17 September at Khormaksar Airport a grenade was thrown in amongst a party of British schoolchildren who were waiting to fly back to the United Kingdom after their summer holidays. Five children were injured, two of them seriously. As a result of this incident and the others which preceded it the High Commissioner dissolved the Aden government and assumed direct rule. In the ensuing riot the security forces arrested some 750 Arabs, many of whom were deported to the Yemen.

In all military operations in Aden, the British Army was bound, as it has been in many other counter-insurgency situations, by the principle of 'minimum force'—that is, in the main troops could only shoot if they had already been shot at. The decision as to what degree of force to use in a particular situation was left to the senior man on the spot—which, in the case of a lone sentry for example, was himself. Written instructions were issued to all ranks (in much the same way as they were later to be issued in Northern Ireland). Provided an individual acted in 'good faith', the security commander made it absolutely clear that he accepted full responsibility for the actions of all the soldiers under his command. With very few exceptions it is true to say that the rule of minimum force was strictly adhered to; indeed there were instances when British soldiers lost their lives when they hesitated to open fire for fear of shooting the wrong man. The principle of 'minimum force' is consistent with the British way of doing things, and although, in a purely military context, it may not be the simplest and quickest way of solving a military problem, British troops in Aden showed a remarkable awareness of the political consequences of the irresponsible use of military force. It is probably true to say that no other army has so consistently abided by this principle in the face of such extreme provocation. In this context it is interesting to compare British tactics with those of the French in Algiers.

By the end of 1965 six British soldiers had been killed and eighty-three wounded. The largest proportion of these casualties succumbed to grenades. In the narrow streets of Aden every door and window presented an opportunity for the terrorist, who could throw or drop his grenade and then slip round the corner and sit nonchalantly in a coffee house with not a shred of evidence to indicate his guilt. Rocket launchers were also used by the terrorists during 1965, though by the beginning of 1966 these gave way to mortars which had greater range and flexibility. If 1965 had been a bad year, 1966 was to be even worse. Though the number of British soldiers killed during 1966 was only five, the total wounded jumped to 218. This was perhaps surprising, because on 22 February the new British government announced that in fact a British military presence would *not* be maintained in Aden after independence. So in theory there was now little to be achieved from killing British soldiers, since they were going to go anyway. But logic does not always prevail.

The decision of 22 February was in line with the new Labour administration's policy of withdrawal from all remaining military commitments east of Suez as quickly as possible. The federal rulers were, of course, shattered by what they understandably regarded as British perfidy. It had been largely the assurance of British military backing that had persuaded them to agree to federation in the first place. On the other hand, President Nasser and the nationalist movements he supported in Aden were much encouraged.

From the point of view of the army the announcement was a disastrous one: what little local support they had enjoyed would now be totally removed, and once again they were put right in the firing line.

On 30 April a terrorist bomb in Sheikh Othman demolished a school bus, killing nine children and injuring another fourteen. It was at this stage that the evacuation of British service families was initiated and a new security commander, General Philip Tower, appointed. At the same time Sir Richard Turnbull was succeeded as High Commissioner by Sir Humphrey Trevelyan, appointed to oversee the British withdrawal from Aden. The stage was therefore now set for the final act of this sordid drama for which the authorities had thoughtfully provided a new cast.

In an attempt to prepare Aden for its independence, the old Federal Regular Army (FRA) had been reorganised into the South Arabian Army (SAA). The metamorphosis was not, however, without its complications. The first rumblings were on 16 June, when a group of senior Arab officers objected to the appointment of a new commander designate of the SAA. Because they did not put their complaint through the proper channels, they were suspended from duty pending an investigation. On the morning of 20 June rumour had it that these officers had been arrested and dismissed by the British authorities; in consequence troops rioted, seized armouries and put their barracks in a state of defence to repel the attack which it was also rumoured the British were mounting against the SAA. The first casualties were nineteen men of the Royal Corps of

Commando and Para soldiers patrol together in Aden town

Transport who were driving innocently past Champion Lines returning from a rifle range practice. They were ambushed at close range by a hail of fire from the barrack huts of the Arab soldiers. Eight British soldiers were killed and another eight wounded; in the ensuing crossfire two policemen, a British public works' employee, and a subaltern of the Lancashire Fusiliers, on duty in the nearby Radfan Camp, were also killed.

Troops of the 1st Battalion the King's Own Border Regiment were ordered to restore the situation in Champion Lines—if possible without firing a shot. Supported by a troop of armoured cars of the Queen's Dragoon Guards, they set out for Champion Lines. Coming a little too close, they were shot at as they dismounted from their vehicles, losing one killed and eight wounded. A platoon of the Borderers immediately made a dash for the SAA guardroom, *without* themselves opening fire, although they were given some covering fire by the armoured cars. Two other platoons then occupied the rest of the camp. Once the SAA soldiers realised that the British were not firing at them, all opposition melted away. A highly complex situation had been solved in a way which perhaps only the British Army would ever have attempted. Major David Miller, the Borderer officer in command of the operation, was awarded the Military Cross, and two of his company were Mentioned in Dispatches.

But the horrors of 20 June were not over. The rumours of an attack by the British had spread to the Police Barracks in Crater. Just at that time Major John Moncur, who was commanding Y Company of the 1st Battalion the Royal Northumberland Fusiliers, had lost contact with one of his platoons; he had ordered them to withdraw from a position near the Police Barracks in Crater and worried about them, decided to go and look for them himself. He and his party were riding in two Landrovers, and he was accompanied by his company sergeant-major, four men of his own company, and also Major Brian Malcolm and two soldiers from the Argyll and Sutherland Highlanders who were soon due to take over responsibility for Crater. As the group approached the Police Barracks they came under heavy fire at almost point-blank range: all except one man were killed. At about the same time, just

after midday, another party of British soldiers narrowly escaped death. The pilot of an army Sioux helicopter was hit in the knee by a bullet just as he was landing a two-man piquet into position on the rim of the volcano overlooking Crater. One of the passengers was knocked unconscious and badly injured in the forced landing, but the other, Fusilier Duffy, pulled both the pilot and his brother fusilier clear of the blazing helicopter before it exploded; he then radioed for help, tended his comrades' wounds and guarded them until help arrived. For his bravery Fusilier Duffy was awarded the Distinguished Conduct Medal.

During the afternoon two attempts were made to re-enter Crater, but both were driven off by intense small arms and rocket fire. It would have been perfectly possible to retake Crater at this stage in a full-scale operation using the 76mm guns of the Saladin armoured cars and accepting the risk of a high casualty rate. But the price would almost certainly have been the complete alienation of the SAA and the Aden Police, which were likely to be the only indigenous organisations capable of running the federation after independence; there would also have been heavy casualties among the civil population. So General Tower was forced to take the militarily unpopular but politically necessary decision not to retake Crater that day.

The 20 June 1967 was a tragic and terrible day for the army in Aden; in all, twenty-two soldiers were killed and thirty-one wounded—yet the number of casualties was no reflection on their professionalism for they had been mown down, not in combat, but by men whom they thought were their comrades. The dead were buried five days later in a blasted stretch of desert called Silent Valley, twenty-five miles from Crater.

Meanwhile Crater was ringed by troops. British snipers methodically picked off anyone seen carrying a weapon illegally, and succeeded in killing ten terrorists in this way. The policy of avoiding civilian bloodshed in Crater at all costs dictated that the area would have to be taken by stealth. SAS patrols therefore slipped into Crater at night, and they were able to confirm that the Arabs were almost without exception off their guard through most of the night. Lt-Colonel Colin Mitchell, commanding the Argyll and Sutherland

Highlanders, and who had just assumed responsibility for Crater, was ordered to re-enter it on the night of 3 July. Although General Tower had intended that it should be a limited 'nibbling' operation, at the very last moment Colonel Mitchell was given permission by Brigadier Charles Dunbar, acting on behalf of General Tower, to push ahead as far as the Chartered Bank building, which was the tallest building in the commercial centre of Crater. In the event he went even beyond this; and what is more, took some newspaper reporters with him. Next morning the Argylls were in firm control of most of Crater, having suffered no casualties themselves and having killed two terrorist gunmen. Their achievement is all the more remarkable in that they had never set foot in Crater before, having only just arrived from Britain. The whole battalion had, however, made sure that every alleyway and building in Crater was firmly imprinted in their mind's eye, from studying scale models which they had painstakingly built in their barracks back in Britain.

After dark on 4 July the remainder of Crater was occupied, including the Armed Police Barracks. Colonel Mitchell was an unconventional man to say the least; he said exactly what he thought, and did more or less exactly what he pleased, and successfully managed to antagonise virtually every senior officer in Aden, particularly Philip Tower, in the process. He certainly captured the imagination of the British people back home, for whom he became an instant hero. Nevertheless, whatever the merits and demerits of his style, he kept a firm hand on Crater for the remainder of the British occupation.

On 13 July the armed police in Crater paraded for and were inspected by General Tower—these were the men who had killed those British soldiers on 20 June. At about the same time, 20 November was announced as the planning date for final withdrawal. So now it was just a matter of hanging on. August was a terrible month, in which there were over 700 terrorist incidents, causing 20 casualties among British troops. Little Aden was handed over to the SAA on 13 September; later in the same month the Paras left the SAA in charge of Sheikh Othman, and also the furious battle then raging in the town between the NLF and

FLOSY. (The NLF had already gained control of the former Aden Protectorate soon after the last British troops had withdrawn to Aden at the end of June. By the 31 August all but one of the states of the federation were under their control, and the federal government had effectively ceased to exist.) By the time the Paras withdrew from Sheikh Othman they had spent four months there, during which they had suffered three men killed and twenty-one wounded. Over the same period they had killed thirty-two terrorists and wounded thirteen. From Sheikh Othman they withdrew to a prepared defensive line which stretched across the isthmus about two miles north of Khormaksar Airfield. This line, known as the Pennine Chain, had been prepared by the Paras and the Royal Engineers during the previous week.

September and October were relatively quiet months for the British as the NLF and FLOSY slogged it out for eventual control of Aden. The British thinned out through November until at last the final scene was played, at 1450 hours on the afternoon of the 29th, when the last company of 42 Commando was flown out by Wessex helicopter to their carrier, HMS *Albion*. Major General Philip Tower was on the last but one helicopter to leave. South Arabia became independent at midnight.

The price paid by the British Army in Radfan and Aden from 1964–7 was 90 lives lost and 510 wounded. Once again the army was required to hold the ring while politicians and diplomats scurried to and fro trying to patch something up. In view of what Aden has since become, it is doubtful if the effort and the cost in lives can truly be said to have been worth it. The young soldiers in Aden faced extreme provocation and continuous danger in very unglamorous circumstances— that they did so with such steadfast and uncomplaining perseverance is truly remarkable. It is a story of which the Army can justly feel proud, a story of vicious urban combat for the highest stakes and in the full glare of the world's media.

Jerusalem 1967

Whilst the British were extricating themselves from Aden, the Israelis were, for the second time

Israeli troops operating in Jerusalem

in the short history of their nation, fighting a war against their Arab neighbours in which the stakes were nothing less than the survival of their country as an independent nation state. Most of the fighting in the Six Day War in 1967 was in open desert terrain, but the battle for Jerusalem involved intense street fighting. The best way to appreciate the nature of this particular urban conflict is to quote verbatim from an account of the battle given by Colonel Mordechai Gur, the Israeli paratroop commander who led the Israeli ground forces in the fighting in Jerusalem.

'On Monday we were ready in the vicinity of our airfields to embark for airborne action [against El Arish]. At about 1400 hours we received the order to take one battalion to Jerusalem, but before long it turned out that a whole brigade was wanted. We were entrusted with a mission which we realised was one of extreme difficulty—to break through into a built-up area. In every army this is considered to be the most formidable type of combat area. We were to connect up with Mount Scopus and create a situation in which we could conveniently break through into the Old City. A few days before this I had gone to look over the terrain, to observe the position, the fortifications, and to take stock of the enemy's position . . .

'We were moving forward with two battalions, one operating in the Police School sector on 'Ammunition Hill', the other in Sheik Jarah. This was fighting of a sort I had never experienced . . . it was going on in the trenches, in the houses, on the roofs, in the cellars, anywhere and everywhere . . . The battle of the trenches lasted from 0220 hours until about 0700 hours. We had reached a bunker with two heavy machine-guns we did not know about, as it was hard to place them from an aerial photograph . . . Just then a soldier jumped up, over the bunker, and dropped a grenade from above, despite the fact he was completely exposed. The grenade exploded but the firing from the bunker kept up. Then one of our men threw three charges of explosives. The first soldier jumped back to the other side of the trench and detonated the three charges. The bunker exploded but only three soldiers were killed and the two survivors continued to shoot. An Israeli soldier then came over from the other side and flung another grenade in—and that was the end of that . . .

'At 0345 the Police School was taken. This was the heaviest fighting of all [40 men were killed out of 500 in the paratroop battalion]. The Police School was held by more than two hundred of the Arab Legion. You should be at least three times as strong as the enemy for an attack against a heavily fortified position. The Arab Legion fought like hell. It took us several hours of street-to-street fighting before we took the school. When we did, 161 Arab Legionaires lay dead in and around the building. The Israeli forces were given supporting fire by 120mm mortars and by artillery in the vicinity of Castel, and two searchlights on top of the Histadrut building were used to illuminate the area so air strikes could continue during the night. . .

'Our two regiments continued to advance, and at about 0600 hours the Ambassador Hotel was in our hands; the whole of the American colony fell to us a little later. Some of the Legionaires retreating from the front line took cover inside the buildings and so there was house-to-house fighting. Sometimes we had to deal with the same house twice, since our boys were running forward all the time. Here we suffered casualties in the streets, for the Legionaires continued to fire from those houses which had not yet been dealt with, and some of our boys were shot from behind. As dawn broke a little after 0400 hours, we engaged our tank battalions, distributing the tanks between the regiments; fighting proceeded to the inner courts while we went mopping up along our main lines of advance up to the Rockefeller Museum. We now threw in our third regiment, the one that had been fighting in the Mandelbaum Gate area . . .'

British, Israeli and French forces were not the only ones engaged in urban combat in the post-World War II scene. The Americans were also reminded of the need to retain their expertise in urban combat.

Vietnam

Perhaps the most bitter battle of the entire Vietnam war took place in the city of Hué, situated not far from the coast and just south of the 17th parallel dividing South Vietnam from North Vietnam. Hué was a lovely old town of temples and palaces, reconstructed by the Emperor Gia Long in the nineteenth century to replicate the seat of his Chinese patron in Beijing. Communist forces crashed into the city in the early hours of 31 January 1968, meeting little resistance from the government division based there. They ran up the yellow-starred Vietcong flag atop the Citadel, an ancient fortress in the centre of town, and then their political cadres proceeded to organise the worst bloodbath of the conflict.

Five months before, as they began to prepare for the assault, Communist planners and their intelligence agents inside the city had meticulously compiled two lists. One detailed nearly two hundred targets, ranging from government offices and police posts to the home of the district chief's concubine. The other contained the names of 'cruel tyrants and reactionary elements', a formula which covered civilian functionaries, army officers, and nearly anybody else linked to the South Vietnamese régime, as well as uncooperative merchants, intellectuals, and clergymen. Instructions were also issued to arrest Americans and all other foreigners *except for* the French—presumably because President de Gaulle had publicly criticised US policy in Vietnam.

Armed with these directives, Vietcong teams conducted house-to-house searches immediately after seizing control of Hué, and they were merciless. During the months and years that followed, the remains of approximately three thousand people were exhumed in nearby riverbeds, coastal salt-flats, and jungle clearings, the victims having been shot or clubbed to death, or buried alive. Paradoxically, the American public barely noticed these atrocities, preoccupied as it was by the incident at Mylai—in which American soldiers had massacred a hundred Vietnamese peasants, women and children among them.

General Tran Do, a senior Communist, flatly denied that the Hué atrocities had ever occurred, contending that films and photographs of the corpses had been 'fabricated'. In Hué itself, a Communist official claimed that the exhumed bodies were mostly Vietcong cadres and sympathisers slain by the South Vietnamese army after the fight for the city. He also blamed most of the civilian casualties during the battle on American bombing. But he did hint that his comrades had participated in at least a share of the killing—resorting to familiar Communist jargon to explain that the 'angry' citizens of Hué had liquidated local 'despots' in the same way that 'they would get rid of poisonous snakes who, if allowed to live, would commit further crimes'. Balanced accounts have made it clear, however, that the Communist butchery in Hué did take place—perhaps on an even larger scale than was actually reported during the war.

Acts of vicious barbarity were widespread. For example, captured in the home of Vietnamese friends, Stephen Miller of the US Information Service was shot in a field behind a Catholic seminary. Dr Horst Gunther Krainick, a German physician teaching at the local medical school, was seized with his wife and two other German doctors; their bodies were found later in a shallow pit. Despite their instructions to spare the French, the Communists arrested two Benedictine missionaries, shot one of them, and buried the other alive. They also killed Father Buu Dong, a popular Vietnamese Catholic priest who had entertained Vietcong agents in his rectory, where he kept a portrait of Ho Chi Minh—telling parishioners that he prayed for Ho because 'he is our friend, too'. Many Vietnamese with only the flimsiest ties to the Saigon régime also suffered.

But it was not just the Northern Vietnamese who acted with such barbarity. South Vietnamese teams slipped surreptitiously into Hué after the Communist occupation to assassinate suspected enemy collaborators; they threw many of the bodies into common graves with the Vietcong's victims. The city's entire population suffered in one way or another from the ordeal.

On 24 February, South Vietnamese troops ripped the Vietcong flag down from the south wall of the Citadel, hoisting the government's red and yellow banner in its place. Many were natives of Hué whose families had been ravaged by the Communists, and they had fought well. But three US marine battalions played the decisive role in the liberation of the city. Myron Harrington, then commanding a hundred-man company, remembered afterwards his apprehensions as a truck convoy transported his unit toward the battle from Phu Bai, a marine base to the south:

'I could feel a knot developing in my stomach. Not so much from fear—though a helluva lot of fear was there—but because we were new to this type of situation. We were accustomed to jungles and open rice fields, and now we would be fighting in a city, like it was Europe during World War II. One of the beautiful things about the marines is that they adapt quickly, but we were going to take a number of casualties learning some basic lessons in this experience.'

Leaving their vehicles, the marines crossed the Perfume river aboard landing craft, with Communist troops peppering the boats from both shores. The weather was cold and clammy, and the low cloud and overcast conditions made tactical air support almost impossible. The marines entered

South Vietnam pre-1975. *Inset*: French Indo-China pre-1945

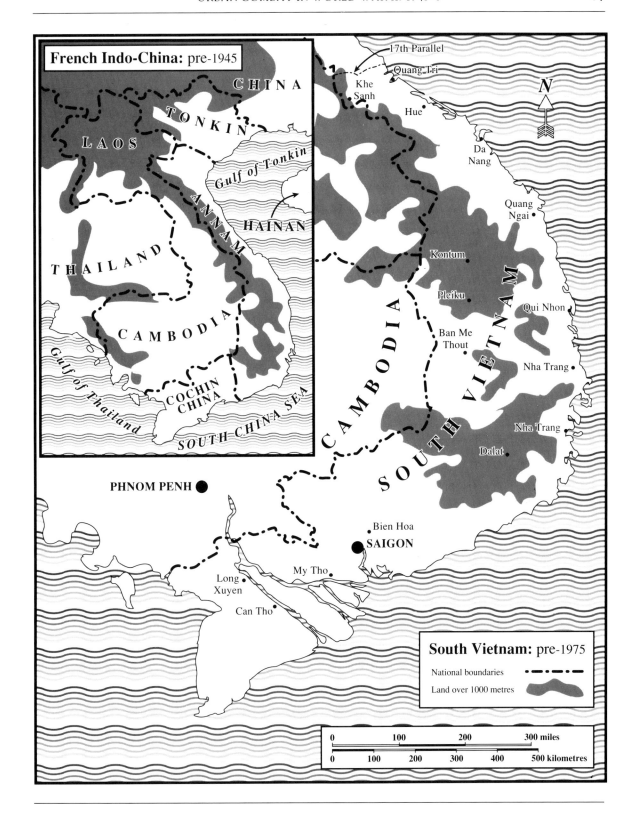

French Indo-China: pre-1945

CHINA

TONKIN

LAOS

Gulf of Tonkin

HAINAN

ANNAM

THAILAND

CAMBODIA

Gulf of Thailand

COCHIN CHINA

SOUTH CHINA SEA

17th Parallel

Quang Tri

Khe Sanh

Hue

Da Nang

Quang Ngai

Kontum

Pleiku

Qui Nhon

Ban Me Thout

Nha Trang

SOUTH VIETNAM

CAMBODIA

Nha Trang

Dalat

PHNOM PENH

Bien Hoa

SAIGON

My Tho

Long Xuyen

Can Tho

South Vietnam: pre-1975

National boundaries

Land over 1000 metres

0	100	200	300 miles

0	100	200	300	400	500 kilometres

Hué from the north, cautiously threading through its streets as they headed toward the Citadel to bolster another marine unit already there.

Harrington's orders, received soon after reaching the marine command post near the Citadel, were to take a fortified tower located along the east wall that was bristling with North Vietnamese and Vietcong troops. The next morning, he launched a frontal assault with two platoons and a tank. Artillery and mortars had shelled the target in advance, but his men faced an implacable enemy. The battle which followed was pure urban combat (unlike most of the fighting in the Vietnam War). The marines crawled and crept into the city; they faced a withering fire from the North Vietnamese regulars and the Vietcong, and had to winkle them out one by one from their spider holes. The marines also made full use of their artillery support, their forward observation officer (or FOO) giving the guns such precise directions that the shells often fell within twenty-five yards of their position inside the fortress. But the battle went on for another ten days as they pushed ahead into the intricate recesses of the old structure, and Harrington recalled that the combat became increasingly close, almost intimate.

'As a marine, I had to admire the courage and discipline of the North Vietnamese and the Vietcong, but no more than I did my own men. We were both in a face-to-face, eyeball-to-eyeball confrontation. Sometimes they were only twenty or thirty yards from us, and once we killed a sniper only ten yards away. After a while, survival was the name of the game as you sat there in the semi-darkness, with the firing going on constantly, like at a rifle range. And the horrible smell. You tasted it as you ate your rations, as if you were eating death. It permeated your clothes, which you couldn't wash because water was very scarce. You couldn't bathe or shave, either. My strategy was to keep as many of my marines alive as possible, and yet accomplish our mission. You went through the full range of emotions, seeing your buddies being hit, but you couldn't feel sorry for them because you had the others to think about. It was dreary, and still we weren't depressed. We were doing our job—successfully.'

During the fight for the Citadel, as in the other battles for South Vietnam's towns and cities at the time, the ubiquitous television crews were present—filming for overnight transmission to millions of American viewers the extraordinary drama of their husbands, sons, and brothers in action.

Seventeen members of Harrington's company lost their lives in the struggle for the Citadel, and nearly 150 US Marines were killed during the entire Hué battle, as well as 400 South Vietnamese troops. An estimated 5,000 Communist soldiers met their death—most of them annihilated by American air and artillery strikes that also inflicted a heavy toll on the civilian population. Urban warfare is appallingly expensive in human lives.

Hué, embodying centuries of Vietnamese history and cultural values, had become a shattered monument after a month of 'house-to-house, room-to-room' fighting. Lieutenant Colonel Harrington's Delta Company had helped lead the assault, yet Harrington himself keenly felt the paradox expressed by so many field officers: 'There was a reluctance initially, of course, to use our heavy armament to destroy the city in order to save it'. He had found the North Vietnamese soldier to be 'an extremely tenacious fighter. He did not flee and run when the marines came in—he held his ground. He was well indoctrinated; he knew what his mission was'.

In fact the North Vietnamese were such tenacious fighters that they proved too difficult to shift without the US marines incurring unacceptable casualties. Thus the decision was taken to bomb the Citadel; and the city was literally blown away. In Hué alone the civilian dead exceeded 5,800, ten times the combined US/South Vietnamese Army losses. But the US army made far too many mistakes. Most crucial of all they failed to isolate the city. Throughout the battle the North Vietnamese managed to maintain a corridor along the line of the Perfume river from the countryside into the city. But most worrying of all, the Americans had allowed their capability in the art of urban warfare to fall into disuse. The marines, traditionally accustomed to close-in fighting in previous wars, had experienced combat in Vietnam only in the rice-paddies and the jungles.

Harrington notes that the marines had not participated in combat in a built-up area since Korea fifteen years earlier ' . . . so that our training was not completely efficient in that area—we do train in that area, but our experience at that time [Hué] was absolutely zero. Initially as we went in we did not have any real concept of how we were supposed to fight'.

And this is perhaps the most important lesson of Vietnam: that armies seem incapable of retaining important skills except by relearning them the hard way.

Beirut

A full consideration of urban warfare in Beirut would necessarily include that unfortunate city's experience over many years since the mid-1970s; in particular it would include an examination of the confrontation between the Muslim and Christian factions in Beirut over a period of time, the Israeli effort to impose a military strategy in Beirut in 1982, as well as the US/French/Italian/British intervention in 1983/84. It would require a whole book to do justice to such a consideration. The most valuable lesson for posterity and for the purpose of this book—how to win an urban battle—can be extracted, however, from Israeli operations in Beirut in 1982.

The multiple crisis in the Lebanon from 1975 to 1982 had long threatened international and regional peace and security. However, in the early 1982, the Palestine Liberation Organisation (PLO) artillery holdings in southern Lebanon increased dramatically, and the PLO force structure was perceived by the Israelis to be more threatening in a number of different ways. On 3 June the Israeli ambassador in London was the target of an assassination attempt by the Abu Nidal group. The Israeli government chose to respond by carrying out bombing raids against Palestinian targets in Beirut and elsewhere. The PLO in turn retaliated with artillery shelling across the border into Israel. This chain of events served as the catalyst to war.

Israeli forces invaded Lebanon on 5 June 1982. By 14 June, Beirut was effectively cut off and the siege had begun. The Battle of Beirut was critical for all those involved. For Israel, PLO forces had retreated north into the city of Beirut. To ensure security in southern Lebanon, Israeli leaders felt it would be necessary to eliminate the Palestinians from Beirut, the centre of planning and coordination for terrorist and military operations, as well as the 'capital' of the Palestinian movement. Beyond these immediate military objectives, Israel's defence minister and chief of staff believed that as a result of the Israeli military initiative, a new order could be established in Lebanon. A government friendly with Israel, strong enough to control Lebanon territory, and willing to recognise and sign a peace treaty with Israel could be established.

Israel's Lebanese allies, principally the Lebanese Forces, saw in the Israeli actions the opportunity to dominate the forthcoming presidential elections. Removal of PLO power and the reduction of Syria's influence would virtually ensure the election of Bashir Gemayel, the Lebanese Forces' commander-in-chief. However, active participation on any significant scale on the side of the Israeli Defence Forces (IDF) could undermine his prospects—such an overt association might diminish his nationalist credentials and preclude Druze, Shi'a, and Sunni electors from supporting him.

The PLO for their part had long recognised that the IDF's Achilles' heel was casualties. They believed that Palestinians familiar with fighting in the built-up area of Beirut could inflict heavy casualties on Israel if the latter entered the city. Once PLO forces were surrounded, the leadership concluded it could reap extremely important political benefits from either a victory (defined in terms of staying in Beirut) or even a prolonged siege.

Syria sought to avoid a full-scale conflict with Israel. Clashes between Syrian and Israeli ground forces erupted only on occasions where Syria was forced to fight—she initiated few if any of these battles, and frequently withdrew rather than fight Israel. The air battles and the destruction of the Syrian SAM system in Lebanon did not threaten extension of the war into Syria; to all intents and purposes, she was only a minor actor in the Battle of Beirut and most of her forces had withdrawn from the city.

Overleaf: Beirut (left); Lebanon (right)

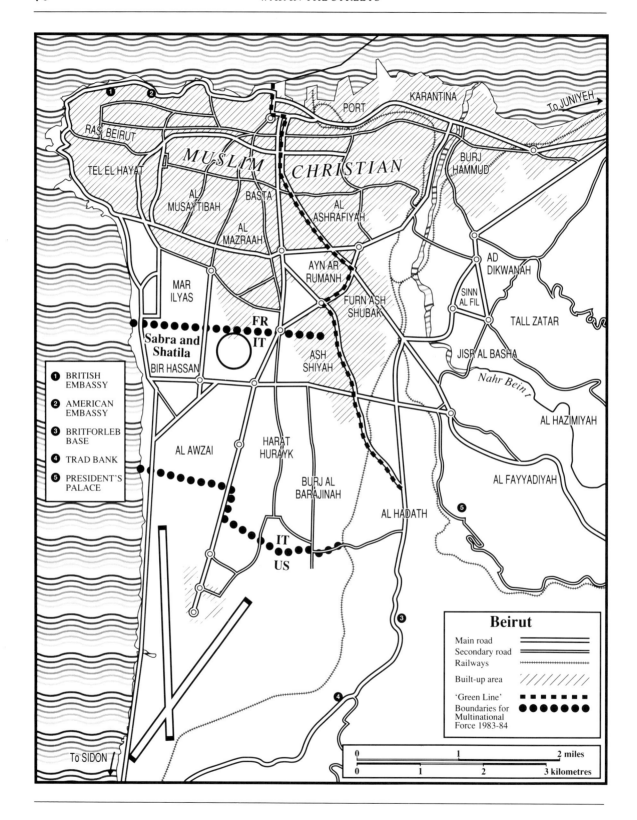

① BRITISH EMBASSY
② AMERICAN EMBASSY
③ BRITFORLEB BASE
④ TRAD BANK
⑤ PRESIDENT'S PALACE

Beirut

Main road	
Secondary road	
Railways	
Built-up area	
'Green Line'	
Boundaries for Multinational Force 1983-84	

0   1   2 miles
0   1   2   3 kilometres

TO JUNIYEH

KARANTINA

PORT

RAS BEIRUT

TEL EL HAYAT

MUSLIM   *CHRISTIAN*

BURJ HAMMUD

AL MUSAYTIBAH

BASTA

AL MAZRAAH

AL ASHRAFIYAH

AYN AR RUMANH

MAR ILYAS

FR IT

Sabra and Shatila

BIR HASSAN

FURN ASH SHUBAK

ASH SHIYAH

AD DIKWANAH

SINN AL FIL

TALL ZATAR

JISR AL BASHA

Nahr Bein?

AL HAZIMIYAH

AL AWZAI

HARAT HURAYK

BURJ AL BARAJINAH

AL HADATH

AL FAYYADIYAH

IT US

To SIDON

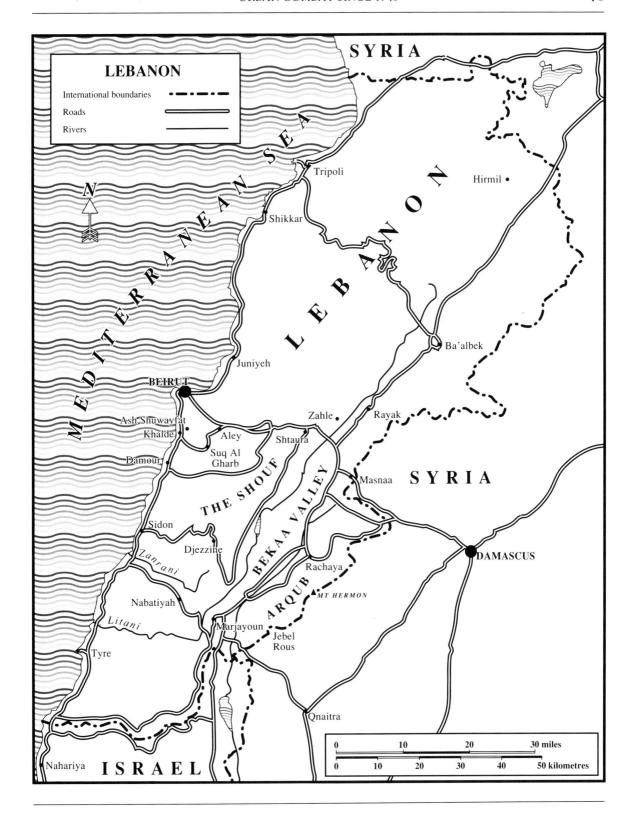

Beirut itself holds about 800,000 people in a city of about 16.5 sq km. The total city and suburban population is slightly over 1,000,000 and the area covers over 50 sq km. Structures in West Beirut north of the corniche reflect the newer building in the capital. Most major construction there postdates 1950, and buildings clearly demonstrate the American influence in architecture. Main thoroughfares have a number of high-rise office buildings, virtually all of framed construction, many using glass extensively. Most of the larger buildings which are not high-rise offices are either apartments or hotels.

The IDF is one of the most highly trained and competent military forces in the world. It is a cadre army, relying on reserve forces to augment its regular units, and is well equipped, highly motivated, and very well led. In the Beirut battle, neutral observer reports reflected the view that Israeli competence and professionalism were highly evident; what set the IDF apart from the other combatants even more than superior equipment was the thorough preparation of individual soldiers and the leadership of the officers. Israeli armed forces soldiers train regularly in live-fire situations and in large unit operations. However, the emphasis on individual initiative and responsibility that typifies the Israeli officer is particularly valuable in urban fighting where *small* units operate with a greater degree of autonomy than in most other advanced armies.

The regular IDF had trained in FIBUA operations prior to the 1973 war. Small unit exercises were carried out in Ismailia (then under Israeli occupation) and Quneitra. After 1978, the IDF carried out exercises in southern Lebanon in villages like Khiam; these exercises demonstrated that the regular IDF was well equipped for FIBUA and that their doctrine was relatively highly developed.

IDF FIBUA training emphasised the coordination of combined arms—infantry, armour, and artillery, along with supporting services, trained together. Reservists however, although highly trained in general, (all having had previous active service often including combat) have only limited annual refresher training—the reserve units therefore were not sufficiently trained in FIBUA prior to the Beirut operation. However, the IDF was saddled with the existing armour-heavy force structure. Also, the dominance of the armour branch within the IDF relegates the foot soldier to a lesser status, with proportionately fewer numbers than are found within other armies—the doctrine for quick, decisive wins in predominantly desert warfare, whilst at the same time minimising casualties, dictates this structure. The small number of infantry accordingly affects the Israeli FIBUA doctrine: less reliance is placed on the role of dismounted infantry than is the case with other armies. This was fundamentally to affect the way the Beirut battle was fought.

Logistics support was also altered for Beirut fighting. Tanks and self-propelled artillery carried food and water as well as ammunition, even though the infantry units to which they were attached also had the capability to carry these items. The support echelons carrying combat supplies moved in armoured personnel carriers (APCs) 100–150 metres behind the front line; medical orderlies moved with the forward troops and carried all necessary life-saving equipment. Despite the unusually developed state of Lebanese medicine, no consideration was given to using Lebanese hospitals for Israeli casualties. Instead, the IDF relied on its own facilities, including mobile surgical facilities, where required.

Collecting stations were established in buildings not exposed to fire, but preferably no more than 150m (or less) behind the spearhead. These stations were light and flexible so they could move with the forces to which they were attached. Those killed in action and the wounded were carried to the collecting station; evacuation from there was by helicopter or ambulance to hospital. Special teams under the command of battalion adminstrative officers were allocated for the evacuation of wounded soldiers at the battalion level; they were particularly trained in communication with and landing of helicopters. The evacuation officers were also assisted by several soldiers of the battalion HQ company specifically earmarked for this task; the soldiers' job was to carry the stretchers with wounded and place them on the helicopters or troop carriers. The battalion doctor maintained an independent station on the battalion communication network, and had his own emergency communications net.

IDF artillery included US standard 105mm, 155mm, and 175mm guns; 135mm guns and captured Katyushka rocket launchers were also used. A large proportion of the shells used were white phosphorous, and considering the limited effectiveness of white phosphorous in cities where construction is generally fire-resistant—which included Beirut—this was surprising. The purpose of this employment is not clear, but it may have been for psychological effect, given the Arab cultural fear of fire and burning.

IDF tanks were of three types—the M-60, the Centurion, and the Israeli-made Merkava. Additional armour plating on the sides and front of many tanks was observed, presumably to confront the problem of anti-tank weapons so plentiful in Beirut. All these tanks fired 105mm rounds.

The IDF introduced some modifications to their equipment holdings for fighting in cities. Additional grenades were carried by a specially designated soldier in each platoon, for example, and pressure charge devices were also supplied. Anti-tank weapons were also provided in numbers beyond the standard issue, as were communication sets. Loudspeakers for psychological operations, not traditionally a part of IDF equipment, were added along with sniping equipment. A proportion of explosive mortar bombs were replaced by illuminating bombs.

Israeli ground forces around Beirut consisted of about 35,000 to 50,000 men; this did not include naval or air forces committed to the campaign, or the ground forces committed to other theatres in the Lebanon war. PLO fighters are thought to have numbered about 6,000 to 7,000 in the Beirut area prior to the war. However, as the war moved toward Beirut, large numbers of PLO fighters entered the city from the south—by the time the city was cut off, there were probably around 12,000 to 15,000 part- or full-time PLO fighters in Beirut. PLO guerillas mingled with non-combatants and, as the fighting neared Beirut, became increasingly careful not to appear in groups without such non-combatants as cover. The PLO deployment in the centre and inner perimeters consisted of small units, mostly of platoon size.

PLO small arms included first and foremost the ubiquitous AK-47/AKM Soviet-made assault rifle, the Belgian FN, German G-3 7.62mm and 12.7mm machine-guns and the RPG-7 Soviet-made anti-tank rocket launcher. The most sophisticated weapons at the disposal of the PLO were some 40 T-34 and a few T-54 tanks, a few dozen BRDM-2 Scout Cars and BTR-152 APCs, BM-21 Katyushka mobile 30-tube and 40-tube multiple rocket launchers, and a variety of recoiless rifles, anti-tank missiles, surface-to-air missiles and air defence guns, and a growing artillery inventory (76mm, 85mm, 100mm, 122mm, 130mm, 155mm). The PLO had 200–400 air defence pieces at the outbreak of hostilities, most of which were mounted on trucks that could move quickly from one position to another. Some recently acquired ZSU 23-4 air defence guns were also available. Abundant ammunition was available in numerous storage areas scattered about West Beirut and the southern suburbs. Ample food, fuel, and medical supplies were also readily available to the defenders. No blockade for less than half a year, even if 100 per cent successful, could reasonably have been expected to affect the military supplies, food, or fuel theoretically available to the PLO.

This was the situation faced by the IDF. At first the Israeli tactic was to lay siege to the city. Their attitude was summed up by General Rafael Eitan on 23 July 1982:

'Time is on the side of the one who knows what to do with it. The terrorists think they are harassing us around Beirut. When we are around Beirut, they are under siege, and it is we who determine the tightness of the siege. If we hit them, they have no replacement for anything we hit, whether it be a cannon or an ammunition depot or a person. We want to turn off the water and electricity. We will not give them fuel, and so we will see who will last longer. If they fire a sniper's rifle, this may be what they are left with, and we will drop a 1-ton bomb on any target we decide upon. We will see who will get closer to a solution finally.'

However, the Israelis did not wish to give the impression of passivity, so further operations were conducted involving a great deal of activity, including the capture of new territory from the defenders. For these, the IDF focused on using overwhelming force on terrain at the periphery of the defenders' control, pushing the PLO into a smaller and smaller enclave. All this served to demonstrate the IDF's willingness and ability to conduct FIBUA operations, and to pay the price necessary to ensure the expulsion of the PLO from Beirut. The IDF opted for a slow advance in order to minimise casualties, choosing to avoid exposing columns to cross-fire. The coastline axis fulfilled this requirement, as the IDF held the territory to its rear and controlled the sea at its left.

The way in which units of infantry, armour and artillery were grouped for a specific operation reflected IDF FIBUA thinking on combined arms. For example, a tank and 155mm SP Howitzer were attached to infantry companies, under the

command of the latter. This force then operated as follows: when in contact with the PLO, the tank was used as a spearhead while the SP Howitzer remained in the rear. The Merkava tank was especially suitable for this approach, as the infantry could use its internal troop compartment, and its armour provided substantial protection against hand-held anti-tank weapons, mines and other light weapons. Exposure of infantrymen to fire and mines was thereby reduced to a minimum. APCs were not used in combat environments as Israeli planners had concluded that they were unreliable as fire platforms, unmanoeuverable and highly vulnerable. In narrow streets, the elevation angles of their machine-guns did not allow engagement of upper-floor targets. APCs were therefore used only in logistic tasks such as carrying supplies for advancing forces, and so endeavoured to remain at least 100 metres behind the front line.

British helicopters evacuating British nationals from amidst the fighting in Beirut. The helicopters landed on the waterfront.

When the advancing force was not engaged by fire, its spearhead was an infantry platoon moving in echeloned columns, each providing cover for the advance of the other. When hostile fire was encountered, the tank moved forward and opened fire. If the source of hostile fire was not located, the tanks fired at suspected hostile positions, especially windows and balconies. If necessary, SP guns moved forward at this time to lay direct fire on targets. Field artillery shells were believed to have a psychological effect on the PLO defenders because of their blast.

For their part, the PLO had not been idle in preparing their defences. These included barricades, tunnels, mines, wired explosives, fortification of basements and other underground levels, trenches and prepared defensive positions. Tunnels were dug between buildings. Headquarters, hospitals and storage dumps were dispersed and often placed below ground in basements.

Deserted apartments were transformed into observation points and weapons positions. The pattern of streets and buildings maximised the vulnerability of traffic, from armoured vehicles to anti-tank weapons. Key buildings were reinforced with extra concrete, sand bags, or anything else to increase wall thickness.

The Israeli siege of Beirut intensified at the beginning of July 1982. On 3 July, Israeli tanks sealed off west Beirut by stationing themselves at the crossing points to the east. Throughout the month of July, Beirut was under siege. This period was characterised by negotiations and intermittent artillery exchanges, as well as small, isolated

Car bombs in Beirut have claimed hundreds of lives. One massive lorry bomb was driven into a block housing hundreds of US marines during the 1983–4 multi-national initiative involving US, French, Italian and British troops in Beirut. It killed 140 US marines

battles. On the 29th, the PLO announced its decision to withdraw its forces from Beirut. But despite this decision, there was little faith in the leadership's sincerity; the Israelis were convinced that the PLO was simply playing for time. The final phase of the battle began on 1 August, with the intensification of the siege and concentrated forward movement by the IDF. Israeli ground forces moved forward to the airport capturing the runways; naval forces used their artillery and Gabriel missiles to attack PLO coastal positions; and the Israeli Air Force carried out numerous bombing sorties. This was the beginning of Israeli 'salami' tactics in Beirut, whereby small pieces of the city were sliced away from PLO control. Significant movement occurred on 4 August, when Israeli forces—which had been reinforcing their positions along the Green Line—launched offensives along several axes from the east and south simultaneously, capturing extensive areas of the city next to the PLO areas. Heavy air raids accompanied the movement; thirty-six sorties on 9 August, sixteen on 10 August, another sixteen on 11 August. There was additional forward movement along the southern coastal axis on 10 August, and the heaviest air raids of the war fell on the 12th—seventy-two sorties. Within a few days, an agreement on PLO withdrawal was reached. The agreement was implemented from 21 August to 3 September, when the last PLO contingent left Beirut.

Detailed Israeli FIBUA tactics warrant close attention, however, because they provide the most up-to-date example of FIBUA in the context of low intensity conventional operations (with the possible exception of Panama City and Khafji), rather than in a counter-insurgency situation such as in Belfast. For example, at staircases two soldiers would move together, one along the wall and the other along the banister. They would leapfrog each other up the staircase in this manner, always covering the turns of corridors and staircases. At the end of the staircase, the two soldiers leaned with their backs against the wall ready to move to adjoining corridors or rooms. Movement along corridors was dependent to a greater extent on the structure of the corridor and the location of adjoining rooms. In principle, movement was along walls, with the first person in the column shouting out the features of the corridor as he saw them. The commander of the group moved in third position.

Building clearing presented a major problem for the Israelis. Their forces had been trained for fighting in built-up areas, but not for combat in cities the size of Beirut, where urban density was extreme. It was recognised that capturing each room in a large building was impossible, and in any case unnecessary, as the PLO avoided close combat and usually evacuated the building. Superficial search and deployment in middle floors of four and five-storey buildings was conducted without fire if possible. If hostile forces were inside and did not withdraw, one tactic was to occupy the second and third floors of adjacent buildings in positions directly facing Palestinian positions. There was a tendency to avoid maintaining a company or any other large number of men in a single building once captured in order to avoid wasting too many men. Once buildings were cleared, forces were deployed at the entrance and on one of the floors overlooking the street.

This, of course, was not the experience of the Canadians at Ortona, of the Russians at Stalingrad, or indeed of the Americans at Hué. When you are faced with a determined enemy in the circumstances of general war, you must hold what you capture, otherwise you will lose what you have gained as quickly as you capture it. However, for the Israelis in Beirut, this was not a problem. Their enemy was of a different quality.

Tanks were used by both sides, but particularly by the attackers—in fact armour is a major ingredient in Israeli urban warfare doctrine. Tanks were used to concentrate firepower on specific targets and to protect the infantry; the IDF believed they were among their most valuable weapons in Beirut, and the most valuable for suppressing hostile fire. Tanks often responded to fire by shooting phosphorous shells into the buildings from which firing originated. Even when the target was not hit, the psychological effect of blast and noise played a role in causing Palestinian fighters to leave their positions. Tank fire was also used to breach walls. Tanks themselves were posted for roadblocks. They played a principal role in the advance of Israeli forces, responding to

PLO harassing fire by moving forwards and firing four to five rounds at the source of the fire. The advance then continued. The Merkava optical equipment was also found valuable, as well as its firepower.

Artillery is also seen as a useful participant in the urban battle by IDF leaders. Israeli experience demonstrated again the very high effectiveness of SP artillery (especially, the US built M109 155mm) using HE in direct-fire roles.

Tactical bombing by Israeli aircraft was often extraordinarily precise and aimed at specific buildings or gun positions. Air-delivered munitions included laser-guided 1,000lb bombs, and conventional free-fall 500lb bombs, cluster bombs, rockets and air-to-ground missiles. Unlike the experience of infantrymen engaged in urban combat in World War II who often found aerial bombing counter-productive, the precision of some of the air-delivered weapons in this war has shed a different light on the use of air power in built-up areas. It was widely reported that a building in which Arafat had been located was totally destroyed by a direct hit from the air, moments after he had left it.

The IDF had 88 killed and 750 wounded in action in Beirut, though these figures include casualties resulting from the fighting along the Beirut–Damascus road south-east of the capital. Predictably, head and neck wounds were found to be particularly common in built-up areas.

From the military viewpoint, it is clear that Israel achieved all its major objectives. The PLO was forced to leave Beirut. IDF casualties were kept to a minimum. A Lebanese president was elected who had cooperated with Israel in the past and who showed every sign of cooperating in the future. Most important, security could be expected to prevail along Israel's northern border. Beirut was also a fascinating example of modern urban combat fought with precision and great expertise by a highly competent army against a second-class enemy.

Panama

The world was taken by surprise by the invasion of Panama by US forces on 20 December 1989. It had been on the cards for some weeks, but the proximity of Christmas had lulled most observers into a state of complacency. Two major urban battles had already occurred in 1989: the first in Beijing in Tiananmen Square in the summer, and the second in Bucharest only a matter of weeks before the Panama invasion. The massacre in Beijing cannot be classified as an example of urban combat in its strictest sense, since it involved the systematic killing of unarmed civilians during a short period of time by regular and heavily armed soldiers. The second most certainly did involve urban combat; for several days the centre of Bucharest saw armoured vehicles, including main battle tanks, slugging it out with fanatical Securitate troops holed up in government buildings. Regular troops sided with armed revolutionaries to overthrow the Ceaucescu regime. Though successful, it was a haphazard and chaotic affair.

The invasion of Panama, however, deserves deeper study for a number of reasons: first, it is the most recent example of the use of FIBUA (in US terminology MOUT—'military operations in urban terrain') by US forces; second, it provides an insight into the methods of solving 'out-of-area' problems; third, it highlights some disturbing deficiencies in US MOUT tactics.

General Noriega, self-appointed President of Panama, had been a thorn in the side of Washington for many years. Not only did he control a country of vital strategic importance to the United States—the US has treaty rights over the Panama Canal, including the right to station troops in Panama—but he was suspected of being involved in drug-trafficking to the United States.

The US invasion took place on Wednesday, 20 December 1989. The initial wave of troops consisted of 2,500 men of the 7th Infantry Division from Fort Ord, California; 1,000 men from the 5th Infantry Division from Fort Polk, Louisiana; and the largest contingent, of 7,000 men, from the 82nd Airborne Division, plus two Ranger battalions from Fort Bragg, North Carolina. Some 2,000 further troops joined the initial invasion force a few days later. The 'opposition' consisted of the regular battalions of the Panamanian Defence Forces who, by and large, chose not to oppose the US invasion; and the so-called 'Dignity' battalions, a motley collection of Noriega loyalists and para-militaries. Although US forces were in control of the country within 24 hours, the 'Dignity' battalions were able to cause a significant amount of mischief for up to a week after the invasion. Five days after the invasion, US troops came under repeated sniper fire in many areas of the city. In several areas the US Army resorted to aerial bombardments. San Miguelito, a slum area on the outskirts of Panama City, was bombed by planes and attacked by helicopter gunships three days into the invasion.

Serious questions have since been asked about US tactics. Was a sledgehammer used to crack a nut? Twenty-six US servicemen were killed, whereas there were 202 civilian fatalities, suggesting that the use of firepower was both overdone and indiscriminate.

US army tactics have always emphasised the use of firepower. This is often appropriate in a general war situation, but is seldom the case in low-intensity operations. There is some evidence to suggest that during Operation 'Just Cause', the name given to the invasion of Panama, US troops were loath to employ basic infantry tactics to winkle out the limited opposition. Firepower was seen as a simple solution. However, this was not only wasted effort but was often ineffective, too. A platoon of infantrymen concentrating automatic fire in the general direction of a single sniper usually serves to confuse the situation and allows the enemy to escape in the chaos. This is precisely what seems to have happened again and again in Panama City. Opposition fizzled out over a period of time, not so much because of the effectiveness of US operations, but because the 'Dignity' battalions felt they had made a token demonstration of opposition to the invaders, and that it would, on the whole, be wiser to succumb to a massively superior force. Seldom were they actually defeated in a straight-fire fight. It is somewhat worrying to note the apparent loss of expertise by US troops in the area of minor tactics and MOUT. But to some extent, this has always been a weakness in US army training.

Operation Just Cause demonstrated very clearly the complete inability of the US military mentality to come to terms with the need to apply force in a measured, efficient and appropriate manner. Even in an urban environment, if an American infantry squad is confronted by minor opposition consisting of small-arms fire, the reaction will not be to use fire and manoeuvre to close with and engage that opposition, but rather to stand off and bring in heavier firepower such as artillery, helicopter gunships, an air strike or a direct-fire anti-tank weapon. The rationale is that casualties will be avoided. In fact this is not always the case. Firstly, there are likely to be greater civilian casualties resulting from the use of indiscriminate firepower; and secondly, the delay in bringing forward heavier firepower means that the enemy is able to cause a nuisance for longer. The longer an action is maintained the more likely it is that casualties will result, particularly if those awaiting the arrival of heavier firepower become less aware of the immediate threat. Whereas an immediate assault on the enemy achieves surprise and a shock effect which may so unsettle him that it will cause all opposition to melt away. This is certainly likely to be the case when dealing with an inferior opponent. In many ways US military thinking tends to invest third-rate opponents with too great a potential to cause damage and provide real opposition. Consequently the art of minor infantry tactics, of fire and manoeuvre and of the surgical application of small-arms fire, seems to have been virtually lost in the US Army—it is an army that has become mesmerised by technology. And unfortunately urban warfare is the form of warfare—with the possible exception of jungle warfare—that lends itself least to the application of technology. Basic infantry tactics are necessary. It is true that the Israelis, against an inferior enemy, were able to use firepower to solve many problems in Beirut—but this may not always be appropriate. An army must at least retain the capability to undertake infantry operations in built-up areas. The US Army appears to have lost that capability.

Another unfortunate example of the US Army's seeming inability to execute efficient urban operations was the invasion of the island of Grenada on 25 October 1983. Most of the fighting took place in built-up areas, and muddle and confusion prevailed. The biggest artillery bombardment of the operation missed the target, Camp Calivigny, the depot of the Grenadian

Panama

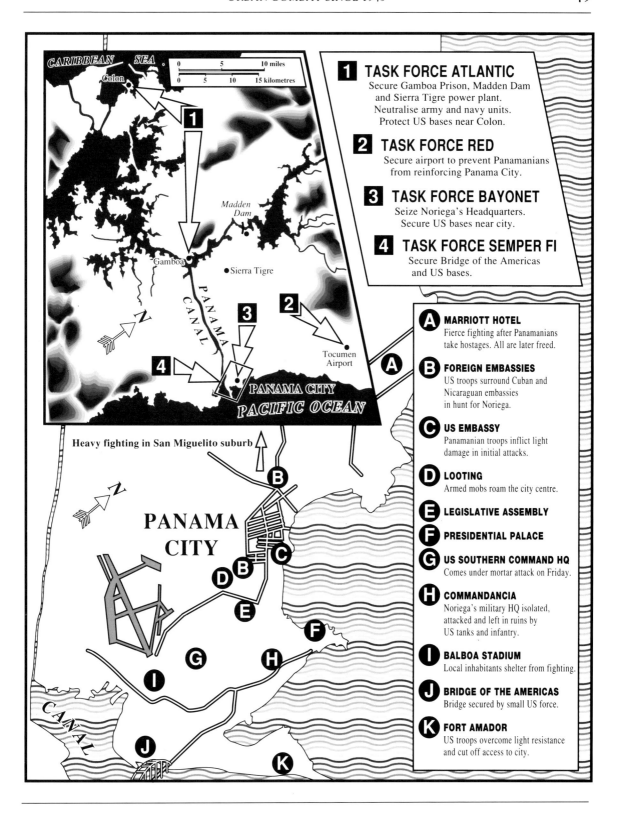

CARIBBEAN SEA

Colon

0   5   10 miles
0   5   10   15 kilometres

1

Madden Dam

Gamboa ● Sierra Tigre

PANAMA CANAL

3

2

Tocumen Airport

4

PANAMA CITY
PACIFIC OCEAN

1 TASK FORCE ATLANTIC
Secure Gamboa Prison, Madden Dam and Sierra Tigre power plant. Neutralise army and navy units. Protect US bases near Colon.

2 TASK FORCE RED
Secure airport to prevent Panamanians from reinforcing Panama City.

3 TASK FORCE BAYONET
Seize Noriega's Headquarters. Secure US bases near city.

4 TASK FORCE SEMPER FI
Secure Bridge of the Americas and US bases.

Heavy fighting in San Miguelito suburb

PANAMA CITY

CANAL

A MARRIOTT HOTEL
Fierce fighting after Panamanians take hostages. All are later freed.

B FOREIGN EMBASSIES
US troops surround Cuban and Nicaraguan embassies in hunt for Noriega.

C US EMBASSY
Panamanian troops inflict light damage in initial attacks.

D LOOTING
Armed mobs roam the city centre.

E LEGISLATIVE ASSEMBLY

F PRESIDENTIAL PALACE

G US SOUTHERN COMMAND HQ
Comes under mortar attack on Friday.

H COMMANDANCIA
Noriega's military HQ isolated, attacked and left in ruins by US tanks and infantry.

I BALBOA STADIUM
Local inhabitants shelter from fighting.

J BRIDGE OF THE AMERICAS
Bridge secured by small US force.

K FORT AMADOR
US troops overcome light resistance and cut off access to city.

People's Revolutionary Army. Worse, when the camp was subjected to a full-scale attack by an entire helicopter-borne battalion, it was found to be empty. Three helicopters and three men were lost in crashes.

When a US brigade was halted by sniper fire from a house, the brigade went to ground and established its headquarters in another house. US aircraft were called up—to deal with a few snipers!—and promptly attacked the house with the brigade HQ in it, rather than the one with the snipers. Five élite infantry battalions took three days to advance three miles and defeat one battalion of third-rate Grenadians and one of half-trained Cubans. Of the eighteen acknowledged deaths of US servicemen, ten were accidental or 'friendly fire' fatalities. Unfortunately, little seemed to have changed between Grenada in 1983 and Panama City in 1989.

This review of just some of the urban conflicts since 1945 has been, of necessity, both selective and brief. Conflict in the Philippines has not been covered; the Soviet invasions of Hungary and Czechoslovakia have not been mentioned, both of which saw fighting in the streets of Budapest and Prague; nor have armed resistance in the streets of Lhasa by the Tibetan people against Chinese occupation; the civil war in Liberia in the summer of 1990 involving street fighting in Monrovia; the deployment of Soviet troops in many of the southern Soviet republics in 1989–90, some of which ended in the use of armed force in cities (such as Tbilisi); the civil war in Sri Lanka, fought to a large extent in urban areas, the overthrow of the Mengistu régime in Addis Ababa in May 1991—all of these, and many more, involved urban combat. There is little doubt that this particular form of warfare is a growing phenomenon, and there are few indications that it is a phenomenon likely to disappear in the foreseeable future.

5   Urban Combat in the Gulf War

The Gulf War was a war by appointment. Thus, after Resolution 678 was passed by the United Nations in November 1990 authorising all necessary means to liberate Kuwait, it was possible to plan in meticulous detail the likely strategic and tactical air and land operations to free Kuwait.

One of those worst-case contingencies envisaged was having to fight street by street and house by house for Kuwait city—mercifully, such a scenario did not materialise. Instead, another totally unforseen urban battle occurred in the early hours of 30 January 1991, when an Iraqi armoured column thrust southwards from its positions in Kuwait and occupied the small Saudi town of Khafji, situated just eight miles from the border. The town had been evacuated some two weeks previously when it came under Iraqi artillery fire; part of the oil storage facility in the town had been set on fire and a pall of black smoke had hung over the town for days. All the world knew that Khafji was empty and unoccupied: images from the deserted town had appeared on television in Europe, in the United States and most certainly in Iraq. US Marine patrols, perhaps 10 men strong, were seen motoring around the town during daylight hours, but it was quite apparent that the town was deserted and, more important, undefended.

The Iraqis, who by this stage had suffered nearly two weeks of unremitting attack from the air and whose air force had effectively been neutralised, had no sophisticated means of gathering intelligence. The allies knew what was happening twenty-four hours a day throughout the Kuwaiti theatre of operations, having at their disposal satellite imagery, photo-reconnaissance aircraft, Remotely Piloted Vehicles, E3 Sentry AWACS aircraft, Sideways-Looking Radar, the new JSTARS system and special forces patrols behind enemy lines—and other systems besides were awash with information about the enemy. The

Iraqis, on the other hand, were electronically blind and deaf. Thus, when presented with the clear statement on television that Khafji was undefended, and needing some sort of success to bolster flagging morale, the Iraqis struck.

So as to remain out of artillery range, the northernmost allied positions were well to the south of the town. A vacuum existed, and Saddam Hussein's forces were quick to fill it. In a lightening move, they dashed southwards from just north of the border and were in Khafji in a matter of hours, certainly too quickly for the allies to make a meaningful counter-move. There was much press criticism at the time of the fact that the allies had been taken by surprise—but it was ill informed. It failed to appreciate that an armoured column can suddenly emerge from behind enemy lines, where it is lost on a radar screen in the 'noise' of all the other enemy formations, and move the 15km to Khafji far too quickly to be stopped if there are no troops actually occupying the ground through which it is moving. Nevertheless, it was a suicidal mission from the outset, but the press, of course, saw it as a great public relations coup for Saddam Hussein. And, indeed, in a way it was: the allies were caught off balance, they were not expecting such a move and it caused some embarrassment.

Once the Iraqi forces, consisting of approximately 2,000 men and 50 tanks, had consolidated in the small border town, they became extremely difficult to dislodge. Any approach across the featureless terrain towards the town was spotted by the Iraqis holding the perimeter of the built-up area, and the Saudi and Qatari troops in the surrounding desert found that they came under intense fire every time they attempted to approach the town. General Schwarzkopf in the Coalition Headquarters in Riyadh took two decisions very quickly: first, the town would be retaken by the Saudi Army with the support of Qatari troops in the area. It was Saudi soil, therefore Saudis must be seen to defend their own territory—this would send an important political message to the Arab world in particular. Second, there would be no unnecessary loss of life amongst coalition forces in the retaking of the town. Militarily, the Iraqi offensive was of no significance whatsoever, and therefore they would be removed methodically and when the moment was right.

The Iraqi force in Khafji was equipped with T54/T55 tanks and BMP armoured personnel carriers. They were in range of their supporting artillery to the north of the Kuwaiti border and, since Khafji was on the coastline, they could be supported or reinforced by Iraqi naval forces; if the Iraqi airforce chose to fly they were well within range of Iraqi close support aircraft. There was, therefore, the potential for causing considerable and prolonged embarrassment to the coalition. But the Iraqi attack was ill-coordinated and was not really pressed home with sufficient support. A force of Iraqi patrol craft and landing craft was intercepted by US and British naval forces heading south along the coast towards Khafji on 30 January. These were engaged by ship-launched helicopters and aircraft, and were all destroyed or beached following the coalition attack. It is probable that this force was attempting to reinforce or resupply the Khafji land forces.

By the time coalition forces had recovered from the initial shock of the attack, they were able to start planning counteraction. The priority was to ensure that the Iraqi bridgehead was not consolidated. British and US aircraft were scrambled to interdict the Iraqi lines of communication between Khafji and the Kuwait border, thus ensuring that no further Iraqi armoured columns entered the battle. British Jaguar ground attack and US A-10 Thunderbolt tank killers in particular took on Iraqi forces north of Khafji. Meanwhile, US Marine Corps 155mm artillery and Cobra attack helicopters completed the ring of steel that was thrown around the small town. The artillery engaged reinforcements attempting to move south and stragglers attempting to move north from the town. The Cobras fired their TOW anti-tank missiles at armoured targets that appeared fleetingly in the town.

But the only way in which the Iraqi force could be defeated was to move into the town, engage them in close quarter combat and defeat them. It took the Saudi and Qatari troops some hours to move into position, but they were able to launch a counterattack late on the 30th, by which time the Iraqis had been in possession of the town for nearly twenty-four hours. Their initial attack had been launched during the night of the 29/30 January. Saudi and Qatari troops used the cover of darkness to gain a lodgement on the Southern edge of the town after nightfall. Since the layout of Khafji is open and the town consists of low, modern buildings, the problems associated with gaining entry to a dense environment did not exist. Moreover, the civilian population had been evacuated, and there was therefore no danger of causing innocent loss of life. As the leading AMX-30 Qatari and M-60 Saudi tanks moved forward, supported by infantry in armoured personnel carriers as well as by artillery fire and attack helicopters, they met small-arms and anti-tank fire, but it was inaccurate and sporadic. A lodgement was achieved relatively quickly, but the Iraqis then proceeded to provide some stiff opposition. Groups of Iraqi infantrymen had to be winkled out of the mostly single-storey buildings. Whilst some caused serious delay to the progress of the coalition forces, many more surrendered, but there were pockets of furious resistance. The Saudis lost several armoured personnel carriers, with 32 wounded and 15 dead, but casualties were much more serious for the Iraqis: they lost 42 tanks, all T54s and 40 dead, many of these in the destroyed tanks. 500 Iraqi prisoners of war were taken. The remainder of the invading Iraqi forces managed to make it back, one way or another, along the coastline northwards to Kuwait, as the corridor they had created was never entirely closed. Although a brigade had launched the original attack, only some 600 Iraqis attempted to hold the town until the end. Khafji was declared clear on 1 February, though it took until the 3rd to eliminate the odd lone sniper who refused to surrender.

One fascinating incident illustrated the cat-and-mouse nature of urban combat. Two US Marine reconnaissance teams, each 6 strong, had been in Khafji when the Iraqis swept in on the night of the 29/30 January. Whether they were so

taken by surprise that they were unable to move out in time, or whether they showed considerable bravery and initiative by staying, is still not entirely clear. We shall probably never know. Whatever the truth of the matter, they remained in Khafji until the town was relieved, sometimes only a few feet from Iraqi soldiers—on one occasion an Iraqi patrol came into a room adjacent to that in which one of the US Marine teams was hiding and remained there for some hours, but for some extraordinary reason did not explore further into the building. Throughout the Iraqi occupation of Khafji, the US Marines continued to direct artillery fire into the town onto targets they could see from their cramped positions.

The burned-out hulks of two Iraqi armoured personnel carriers lie where allied firepower halted their advance during fierce combat in Khafji (Associated Press)

The battle to retake Khafji was initially quite intense, as the 15 Saudi deaths and 32 wounded indicate. In addition, 11 US marines were killed on the first day of the Iraqi incursion by a round fired from an Iraqi T55 tank, this incident taking place in the border area north of Khafji. But the combined weight of Saudi and Qatari armoured infantry and tank forces, supported by US artillery and attack helicopters as well as US A-10 aircraft and British Jaguars, induced after only a few hours of fighting a general collapse of Iraqi resistance, as is indicated by the surrender of 500 Iraqis.

Khafji was a significant urban battle, but it could have been a great deal bloodier if the Iraqis had decided to fight for longer. One of the key factors in their relatively rapid collapse was the lack of any air support whatsoever; of equal importance was the fact that Iraqi artillery support dedicated to the Khafji operation was neutralised by US counterbattery fire early in the operation. The Iraqi troops were therefore very quickly isolated and the battle thereafter was not fought on equal terms. Nevertheless, Khafji demonstrated again how relatively difficult it is to remove entrenched troops from a built-up area. It also demonstrated that urban combat has a publicity value: the fact that Saddam Hussein had managed to seize a Saudi town was judged by the world's media as infinitely more significant than a battle which might have taken place in open desert. Allied competence was questioned, and there is no doubt that the affair was a considerable public relations coup for the Iraqi régime.

There remains one final question about the Gulf War. What would have happened if the Iraqis had decided to fight for and in Kuwait City? There is little doubt that it would have been a protracted and bloody struggle. Coalition forces would have been extremely reluctant to get involved in such a battle and would almost certainly have opted to starve out Iraqi forces, cutting off water supplies and at the same time hoping that the civilian population would either quit the town or remain unharmed. But such hypothetical questions serve little purpose: Kuwait City was liberated without a shot being fired.

The Gulf War showed that even in the context of a desert war, the urban battle can assume a high profile. And methods in Khafji were essentially no different to those used in Ortona or Cassino, Goch or Berlin. Urban warfare changes very little.

6 Contemporary Urban Combat: Offensive Operations

The general principles of attack and defence hold good for operations in built-up areas. How these principles are put into effect depends to a large extent on whether operations are taking place in a small village, a collection of farm buildings or in a large city. Whichever it is, it will always be primarily an infantry battle. Tanks and field guns can be devastatingly effective in some circumstances—as the Israelis demonstrated in Beirut—but are often impossible to manoeuvre or to bring to bear in city streets. In the final analysis, the infantryman is nearly always required to finish the job by fighting at a very low level, that is at platoon or even squad strength. To implement company or battalion operations in the streets of a large city would be very difficult.

Conventional warfare and FIBUA are markedly different in a number of ways:

1 Fighting will be at very close quarters. The enemy will be a house or a street away, sometimes only a room away or just the other side of a wall. Hand-to-hand combat may be required.

2 It is always difficult to locate the source of enemy fire in built-up areas. The 'crack' and 'thump' of a high velocity bullet echoes and re-echoes off surrounding buildings so that it is virtually impossible to tell where it has been fired from. But not only is it difficult to hear where a round has been fired from, it is also usually extremely difficult to locate a sniper visually. A well drilled defender will take maximum advantage from the urban landscape and site his weapons well back from windows and doors. Further confusion can be added by the smoke and dust which inevitably collects and hangs in the street.

3 Fields of fire and observation are much more restricted than is usual in conventional operations. The attacker must expose himself in order to make progress. This leaves him prey to snipers who are particularly effective in an urban environment. In open countryside there are likely to be a number of alternative approaches towards an enemy position. The difficulty of urban combat is that in the urban situation there may only be one line of approach, and your adversary knows that as well as you.

4 Tanks can give very effective close support but, unlike operations in open countryside, they must be protected by infantry. If a tank ventures along a street that has not first been cleared by infantry, it is likely to be engaged by anti-tank fire from close quarters from the side or rear, where its armour is thinnest.

5 A characteristic of fighting in built-up areas is the appalling quality of VHF radio communications. Although remote antennae can be placed on top of buildings, reliable communications at squad and platoon level are very difficult in a dense urban environment. Even in contemporary warfare, infantrymen could be reduced to using hand or light signals, or communication cords.

6 More than in any other category of warfare, there are likely to be civilians in the midst of an urban battle. Their presence obviously complicates the execution of operations.

Just as the circumstances in which an urban battle is fought are markedly different, so also are some of the ground rules of urban combat. First, because of the complexity of the task, **battle plans** must be simple and progressive; a step-by-step approach is usually best. During the course of an operation, platoons or squads are unlikely to be able to work as a unit; they are much more likely to be working in pairs or even as individuals. Thus there is often only the one opportunity for briefing, at the outset of an assault. Then events in urban combat very often take on a momentum and direction of their own—unlike in open countryside, when they can be shaped or influenced more easily in the course of an episode. Thus in urban combat, only the simplest plans are likely to survive the test of battle.

Second, commanders must be prepared to **delegate control** to a greater extent than they might otherwise. Simply because fields of view are so limited, it is clearly more efficient and more reliable for commanders to have control over smaller areas. Thus company commanders, for instance, must divide their area of tactical responsibility into sectors giving platoons and squads limited objectives within these sectors.

Third, urban combat calls for a much greater degree of **thoroughness**. Unlike the rural situation, where the maintenance of speed and momentum may be more important than clearing every copse and fold in the ground, it is a different matter altogether when clearing buildings. Every room, cellar and attic must be cleared, checked and rechecked. It is a dangerous and painfully slow business, but the only way to get the job done.

Houseclearing

The basis of all offensive operations in urban combat is houseclearing. This requires meticulous preparation. Each soldier must shed any bulky equipment such as packs, bergens and entrenching tools so that he can fit through narrow holes, passages and doors—and move as fast as possible in any circumstances. Assault equipment such as toggle ropes, grapnels and aluminium ladders are vital, and torches and binoculars useful. Each man will need additional scales of rifle or machine-gun ammunition, grenades, and as many as possible should carry a grenade launcher. Tracer is indispensable. In the confusion of the urban battlefield, the simplest and most efficient method of indicating a target is by firing tracer at it. Each assault unit must have some method of blowing holes in the sides of buildings in order to gain entry—this could be an anti-tank weapon or a prepared frame charge. Most important of all for morale, careful medical preparations must be made. Casualties are likely to be high in urban combat, therefore extra shell dressings and morphine should be carried by each soldier. Extra

drinking water for casualties should be taken as far forward as possible, and extra stretcher bearers will need to be kept in reserve. Commanders should make detailed plans for the evacuation of casualties before committing their men to battle.

The necessary preparations made, the dangerous business of houseclearing can be undertaken. A squad is about the right size to tackle an average-sized house. Platoons and squads can help each other by working in parallel, for instance, down each side of a street. Precisely how a squad should be organised to clear a house depends upon the circumstances in a particular situation, but the standard formation is to divide the squad into two groups, namely the assault and the covering groups: the first consists of the squad leader, two entrymen, two bombers and one lookout or linkman; the second consists of the squad second-in-command and a machine gunner. If the squad is larger than eight men, there is the option of reinforcing either group depending on the tactical situation.

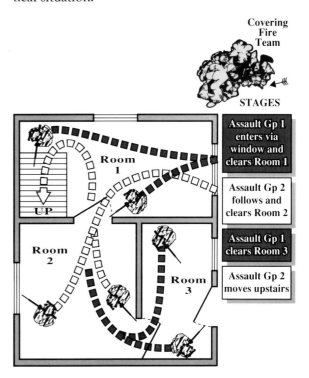

(Opposite, above) Infantrymen assault through smoke and dust around the corner of a house

(Opposite, below) Soldiers preparing to place a frame charge against a wall, to gain entry to a house

Section house clearance operation from ground floor entry

The drill is as follows:

1   The covering group should first take up a fire position to cover the point of entry and, if possible, pre-empt any enemy trying to escape.

2   As soon as the covering fire is effective, the assault group approaches the building and the bombers lob in two or more grenades. The entrymen then enter the house on the heels of the explosion before any survivors have recovered from the shock of the initial assault. It is always preferable to enter a building from the top storey and to work downwards. As a general rule, a house should never be entered from the ground floor. It is likely that a hole will have to be blown in a wall to gain entry. Once the entrymen are inside the house, they clear the room they have entered and then indicate to the squad leader and the remainder of the assault group that it is safe to follow.

3   The members then clear the house room by room, storey by storey, from the roof to the cellar. The lookout/linkman remains by the original point of entry and keeps contact with the covering group.

4   Finally, when the house has been cleared, the squad leader reorganises his men inside or outside the house depending on his next task.

In addition, there are certain ground rules which should be observed when clearing a building. It is always wise to throw a grenade into a room before entering it. Immediately it has exploded, the room should be entered and any surviving enemy engaged with bullet and bayonet. Rather than wait for a reaction, entrymen should fire into cupboards and other likely hiding- places. It is also a wise precaution to shoot into the ceiling and the floor to discourage any enemy who may be above or below. Finally, if it is necessary to move upstairs, it should not be without fire support, ideally from a machine-gun sited outside.

The urban landscape is unforgiving. There are no folds, there are no patterns, few bushes and little undergrowth into which the soldier can melt when he feels threatened. Fighting in towns requires a degree of commitment which is not usually required in a rural setting. In the countryside there are usually a number of routes to an objective; the enemy is not always sure which way

Sequence for mounting a multi-storey staircase

you are coming, and even if he is, it is not always possible to spot a man if he is camouflaged well. But the stark angles and corners of a townscape throw the human form into obvious relief; the soldier feels more vulnerable; there is less cover. You cannot creep unseen around a corner of a building. You have to commit yourself to rush around it in a determined act of courage. The same applies if you have to assault a house: you cannot creep into a room—you must charge into it. Thus the psychology of fighting in built-up areas is altogether different.

In the countryside, an infantryman is taught a different way of fighting—how to make use of every fold in the ground, every blade of grass to hide himself and to get himself by leopard crawl or other approved method from point A to point B. However, many of these techniques are just not suitable for urban warfare. Of course alternative techniques have been developed, all of them making maximum use of the principles of fire and manoeuvre, but all of them leave the infantryman more vulnerable. There is no doubt about it. Urban warfare is dangerous, frightening and difficult. And the pressures on leaders to convince their men to undertake this form of warfare with commitment will be more than usually demanding.

This somewhat clinical description of the mechanics of houseclearing pays scant heed to the terrifying realities of such an operation. Houseclearing in contemporary urban warfare can be physically exhausting, mentally demanding, immensely stressful and enormously frightening. Those engaged in such operations would be unlikely to be able to manage it for long.

Clearing a Street or Village

Houseclearing techniques form the basis for larger and more complex FIBUA operations such as clearing a street or village. Clearing a group of buildings is essentially a combination of a series of house-clearing operations. Faced with clearing a typical west European fairly wide street, the optimum solution is probably to use two platoons, one on each side of the road. The advance should be controlled by the company commander who should keep his third platoon in reserve to deal with unforeseen eventualities. It is sensible for the two platoons to 'leapfrog' down the street, so that one platoon is ahead of the other at any one time; then it can fire straight across the street at any

enemy whom the rear platoon is about to attempt to dislodge. Within each platoon, squads should support each other. Having captured one house, a squad can turn it into a firm base, from which another squad in the platoon can mount its attack on the next house. Of course holding each successive house, as a platoon advances down a street, presents a considerable manpower problem, although this does depend on the nature of the enemy. The problem faced by the Israelis in Beirut in 1982 was very different to that faced by the Canadians in Ortona in 1943. In Beirut, PLO resistance was transitory and fleeting—the enemy did not attempt to reoccupy empty buildings once they had been 'moved on'. Whereas the German paratroopers in Ortona were in a different league altogether: if a house was not fortified and occupied after it had been taken by the Canadians, they would infiltrate back into it.

A determined enemy can therefore make urban combat an extremely expensive undertaking in terms both of time and manpower. This raises the

A building prepared for defence

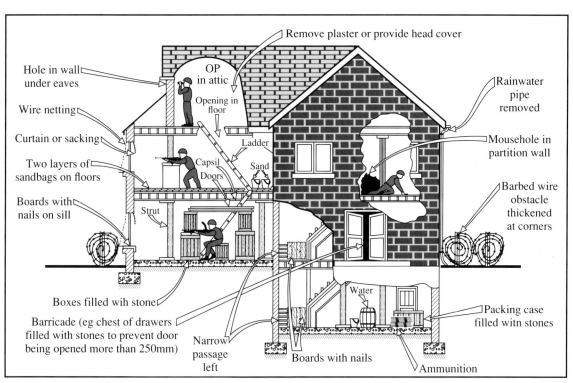

Remove plaster or provide head cover

Hole in wall under eaves

OP in attic

Opening in floor

Rainwater pipe removed

Wire netting

Curtain or sacking

Ladder

Mousehole in partition wall

Two layers of sandbags on floors

Capsil Doors

Sand

Boards with nails on sill

Strut

Barbed wire obstacle thickened at corners

Boxes filled wih stones

Water

Packing case filled witn stones

Barricade (eg chest of drawers filled with stones to prevent door being opened more than 250mm)

Narrow passage left

Boards with nails

Ammunition

The obvious way into a house is through a window—perhaps too obvious

whole question of the wisdom of bypassing a town or village in the interests of the maintenance of momentum, of which more later. However, one solution is to identify and hold the dominating buildings in the area so that surrounding buildings can be covered by fire and thus isolated. Such a task can be allocated to individual snipers, thus conserving manpower. It is, of course, more difficult to sanitise buildings at night, despite the introduction of Night Observation Devices (NODS) and Image Intensification (II) Equipment. It is unrealistic to expect lone snipers to gaze through II sights all night long.

If access through ground floor windows is blocked, it may be necessary to use ladders to gain access through first floor windows

Clearing Larger Villages and Small Towns

Larger villages and small towns can soak up at least a company, and more probably a battalion. Towns and cities could involve brigades or even divisions. The potential of an urban area to impede, delay, complicate and frustrate progress is enormous. Urban areas are 'force multipliers', in the military jargon—in other words, a few determined and properly equipped men can hold up a much larger force for a disproportionate period of time. Techniques for clearing a village or town follow similar principles to urban operations at a lower level. First, a cut-off group should position itself so as to be able to block the enemy's retreat. Second, a covering group or groups should cover the main thoroughfares in the village and engage any enemy soldiers they can positively identify. Third, the assault group or groups should then systematically clear the village. This may well mean a house-to-house battle.

Cities

A city the size of, for instance, Hanover or Braunschweig in Germany could soak up several divisions of troops. Stalingrad and Berlin during World War II demonstrated this tendency very clearly. More recent examples are Beirut in 1982 and Panama city in 1989. Thus, if bypassing is an acceptable tactical option, it should not be disregarded out of hand. However, the dangers of bypassing are illustrated by the example of Bastogne in 1944 (see p.40); this island of resistance and firepower provided both a focus and a *raison d'être* for an Allied counter-attack. As a general principle, islands of resistance should be nipped in the bud, otherwise they can become a thorn in the side. Soviet doctrine for the offensive, however, lays great emphasis on keeping to set norms of progress. The Soviet leading echelons could be expected to bypass large cities put in a state of defence, leaving second echelon formations to take on the task of clearing urban areas. NATO doctrine is not quite so explicit in this respect.

In the foreseeable future, urban combat is much more likely to occur in the context of low intensity conflict rather than general war in Europe; there may well be more urban battlefields like Hué, Beirut, Bucharest, or Panama City.

Urban counter-insurgency also looks set to continue in Belfast and more sporadically in parts of Central and South America, Spain, the Indian subcontinent and in some of the southern Soviet republics. The techniques of offensive operations in an urban environment therefore need to be kept alive. The British Army, due mainly to its experience in Belfast and Londonderry, has developed a high degree of expertise in low-level urban tactics. The Israeli Army, due to its experience in Beirut, is well versed in the complexities of urban warfare at battalion, brigade and even divisional level. The latter is perhaps the only contemporary army recently to have undertaken all arms operations involving artillery, engineers, armour, helicopters and aircraft as well as infantry in a majority. The US Army, however, seems unable to extricate itself from its persuasion of seeing firepower as the answer to every situation in urban conflict. In fact the opposite is true; although a stubborn enemy often requires a shell from the main armament of a tank (if one can be brought to bear) or from an anti-tank weapon to dislodge him, infantry must still manoeuvre in order to bring these weapons to bear; there is no substitute for infantrymen fighting their way into houses and then from floor to floor and room to room within those houses. What is perhaps surprising is how little contemporary urban combat has changed since World War II. Minor tactics are important; the skill of the individual infantryman is still relevant; the rifle, bayonet and grenade still count.

The catch-phrase 'the urban jungle' is often used to describe the environment in a large city, and it is a particularly apposite metaphor as regards urban combat. Precisely the same category of skills is required for FIBUA as for fighting in the jungle: individual initiative, personal courage, low cunning, a high level of marksmanship, patience and supreme physical fitness. Firepower has its place, but the old individual skills of the infantrymen are still vitally important in urban combat.

(Opposite, above) Infantrymen entering via windows

(Opposite, below) But, by far the best, is to get onto the roof thereby gaining the advantage of height

A British infantryman moves through an urban training area in Cyprus. He is carrying a general purpose machine gun (GPMG)

An infantry squad moving down a village street. Here they stop to await further orders. When stationary they must cover the windows above them, the street in both directions and any other potentially threatening target

7 Contemporary Urban Combat: Defensive Operations

Urban areas provide the ideal defensive bastion. Western Europe has become increasingly urbanised in the past thirty years, particularly in Germany where the German economic miracle has accelerated the pace of industrial growth. Before the revolutions of 1989 the Soviet threat to Western Europe was a very real one, and NATO defensive plans took maximum advantage of the many built-up areas in close proximity to the border with East Germany. However, the possibility of a Soviet attack on Europe is now so remote that it can be virtually discounted. Yet central Europe, the Soviet Union itself and the Balkans in particular remain unstable and potentially explosive areas; nor does it take much imagination to visualise other potential trouble-spots worldwide. There is every reason therefore for professional armies to maintain the art of putting a house or a town in a state of defence.

The Defending Force

A force defending a built-up area should be divided into a perimeter force, a disruption force, the main defensive force, and the reserve. The **perimeter force** consists of a number of separate reconnaissance forces whose task it is to establish outposts on the perimeter of the built-up area to cover the most likely approaches. Specifically their tasks will be to give warning of the approach of the enemy, to engage and if possible to destroy enemy reconnaissance and leading elements and, finally, to force the enemy to deploy and mount a deliberate attack in order to break into the town. A perimeter force would normally operate in small groups of perhaps squad or platoon strengths. They would be armed with a high proportion of automatic weapons and anti-tank weapons. If possible they should be highly mobile, in agile reconnaissance vehicles. The job calls for strong nerves and a fine sense of timing: retire too soon and little will have been achieved, leave it too late and risk being cut off. It would be the responsibility of the perimeter force to extricate itself in a timely fashion and get itself back into the relative security of the town without suffering unnecessary casualties.

The task of the **disruption force** would be to cover the ground between the perimeter force and the main defended localities in the heart of the town. Its aims would be to delay, confuse and disrupt the enemy. Delay can be caused by snipers and tank-hunting teams; also the creation of artificial obstacles is clearly one of the most efficient methods of imposing delay—rubble, overturned buses and cars, or telegraph poles. Terror and confusion can be sown by laying mines and planting booby traps. It could conceivably happen that the disruption force is so successful that it discourages the enemy from entering the city centre. In fact it is more likely that firstly it will aim to cause maximum casualties, but its main objective will be to try and channel the enemy—by clever use of obstacles and deployment of forces—into the defended localities where the defensive force can trap and destroy him. The only roads that should be left unobstructed are those which channel the enemy into the chosen killing ground. All other routes should be blocked either by blowing buildings across them or constructing roadblocks.

The main battle, however, will be fought by the **main defensive force**, which should be located in strong positions in the heart of the town with tank support if possible. This is where the enemy is fought to a standstill. This is where the defensive force creates its strongpoints and this is where it stays.

The final constituent part of an urban defensive force is the **reserve**. It is designed to seal off and destroy any penetration. If a strongpoint is overrun, it is the task of the reserve to counter-attack and to restore the position.

This, in outline, is how an urban area is

defended, but it is the detail of the urban battle that is important. More than any other category of warfare, urban combat is a series of individual actions. The mechanics of how to fortify a strongpoint and how to fight from it bear detailed study.

Selecting a Strongpoint

It is important to select the right building to fortify. If it is too small then a single hit on the building from enemy artillery or a tank could well kill all the defenders. On the other hand, too large a building may force a commander to spread his defenders so thinly that he will be unable to cover all the approaches or provide an adequate concentration of fire to prevent the enemy storming the building. Selecting the right building is probably the most difficult decision, and size is not the only criteria. It must also be of the most suitable construction. Old farmhouses and village houses in Europe are often timber-framed with a daub or brick infill; these are inflammable and easily reduced to rubble, particularly by a tank's main armament. Equally, modern bungalows and small two-storey houses are built in part from materials such as plywood or light brick, and do not provide protection against even small arms fire. High rise buildings, the basic framework of which is made from steel or reinforced concrete infilled with large areas of glass, are not particularly suitable for defence, either. Although they are excellent for observation purposes, they are prone to progressive collapse if the building is damaged on the lower floors.

There is little doubt that the more traditional masonry house with strong walls made of brick or steel and probably three or four floors high, is more suitable for defence. This type of house is usually pre-1940, has relatively small windows, is much less flammable and has—certainly in Europe—good solid cellars. Its more modern version, built of brick or concrete block, with a roof and concrete ground and upper floors, is also suitable for defence.

The surroundings of a house also affect its selection as a strongpoint. Although open fields of fire from the house are desirable, it should be neither isolated nor overlooked. Covered approaches to and from the building should be

available for the purposes of reinforcement and supply. Some space around the building allows the defender to lay mines and construct booby traps around the house to prevent the enemy getting sufficiently close to place charges against the building. It also allows the defender to engage his enemy early and effectively.

Preparation

The next task is to increase the protection factor offered by the building. Walls can be strengthened by placing sandbags against them; earth-filled cupboards, chests-of-drawers and mattresses provide additional protection. All entrances should be barricaded, and staircases and passages blocked. Movement around the building for the defenders can be achieved through holes in ceilings which are reached by rope ladders or even piled up furniture. However, this sort of treatment imposes additional pressure on floors and ceilings, and collapse resulting from the shock of an explosion can be avoided by shoring up ceilings with strong timbers. Finally, there should be as much water as

Loophole construction

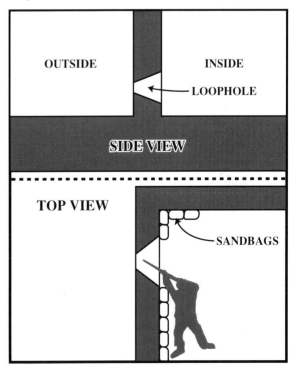

An infantryman with a SA–80 rifle takes up a defensive position

possible stored in basins, buckets and baths ready for use in the event of fire. Gas and electricity should be switched off at the mains.

Siting Defenders and Weaponry

Once the building is prepared, the next task is to site the defenders and their weapons. Automatic weapons should be put near ground level to maximise the 'beaten zone' effect of a machine-gun—these weapons cover a long elliptical beaten zone if the rounds are travelling parallel to the ground; thus their potential coverage is greater than if they are sited firing downwards. Moreover such beaten zones can be covered at night, because their coordinates can be recorded in daylight by means of the dial sight which is fixed to many contemporary automatic weapon systems. Thus, in the manner of artillery fire, areas or targets are recorded and can be dialled up in darkness or bad weather if a threat is noted in that area by a sentry or seen fleetingly through night vision equipment.

Snipers, on the other hand, normally engage lone targets and are better sited higher up so as to maximise visibility. Similarly hand grenades are

Dominating the rooftops is important for defence. However, in this situation infantrymen are vulnerable to mortar, artillery and air attack

better thrown from first-floor level or higher. Hand-held anti-tank weapons are usually more effective in upper storeys so they can engage the top armour of tanks; the British LAW 80, the Soviet RPG-7, and other hand-held anti-tank systems are ineffective against the front armour of most tanks. The main complication with a recoilless anti-tank weapon is that its back-blast makes it unsafe to use in all but the largest rooms—there must be sufficient space behind the firer for the back-blast to be dispersed. The PLO managed to fire RPG-7s from inside buildings in Beirut, and the IRA have been known to do the same in Belfast, but extreme care is necessary.

There are certain techniques to which infantrymen must adhere in defensive urban warfare if they are to maximise their advantage. It is a mistake, for instance, to take up a fire position in a window space. We all know that it is relatively easy to spot someone who is standing close to a window. However, if he steps back perhaps only two metres he becomes invisible to an observer outside. This is merely a function of the relative light levels inside and outside a house. The disadvantage to taking up a firing position back from a window is, of course, that the arc of fire is drastically reduced. The optimal solution is to construct loopholes in unexpected places such as underneath a window sill or through the tiles in a roof. This not only allows the firer to be close to his cover and therefore able to see further, it also provides an additional degree of protection and camouflage. Double bluff is, of course, always an option: dummy loopholes can be made indistinguishable from the real ones. A loophole should be constructed so that the aperture on the outside of the building is small, with a wider section on the inside, in other words V- or cone-shaped. This has the obvious advantage of allowing maximum movement inside the building thus allowing the firer to bring his weapon to bear over as wide an area as possible, whilst at the same time minimising the opportunity for an attacker to acquire his target. To avoid splinters, firing positions should be surrounded with as many sandbags as possible. Any glass should be removed from the windows and, if the time and the stores are available, replaced with anti-grenade wire netting. Even simple precautions can be effective, like

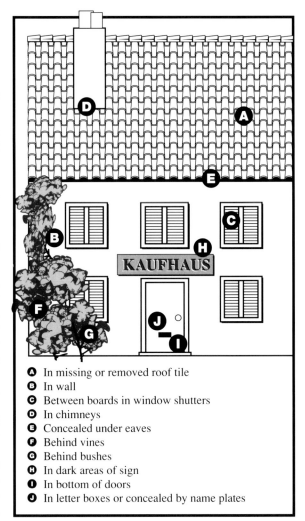

A In missing or removed roof tile
B In wall
C Between boards in window shutters
D In chimneys
E Concealed under eaves
F Behind vines
G Behind bushes
H In dark areas of sign
I In bottom of doors
J In letter boxes or concealed by name plates

Siting loopholes

using net curtains to prevent the enemy seeing into the house.

Defending a house or a collection of houses will often be a lonely task. A soldier in a rural environment can often see for miles—he will certainly be able to see his company area, perhaps even parts of his battalion area. In areas of north-west Europe, where the hilly terrain affords views over long distances, it may even be possible to see the positions of other battle groups in the brigade. The soldier therefore feels part of an unbreakable line; he is part of the environment in which he and his comrades are operating. However, stuck in the middle floor of an office block in a modern city he is likely to feel isolated and vulnerable. For long periods of time he may not even be able to see members of his own squad in the same building, let alone soldiers in other units. Holding a large urban area requires a high degree of decentralisation and delegation, and squad and platoon commanders must often work cut off and with the minimum of direction from their commanders. Similarly individual soldiers will need to use their initiative and take important decisions—such as whether to open fire or not—on their own.

In a conventional defensive position in a rural situation the squad commander will be able to see all of his squad, and it will be he and he alone who will give the order to open fire or not. This is not the same in defensive urban warfare which requires a high level of individual training and also patience and discipline. Commanders must make their intentions absolutely clear from the outset, and thereafter they must trust their subordinates to work within the guidelines or directives which they have laid down.

Tanks, Artillery and Engineers

The successful defence of an urban area can be enhanced by the sensible use of tanks, artillery and engineers. **Tanks** should form an integral part of the perimeter force and the mobile central reserve. They can also be useful in defended localities, providing mobile support for strongpoints or in the anti-tank role from prepared fire positions. On the other hand if they are used as static pillboxes in the open, they are likely to be isolated and eliminated, since a tank loses its natural advantage as soon as it is prevented from using its mobility. Thus in the context of the defence of a large city by a battle group, brigade or even division, armour should be kept in reserve to operate en masse to provide the sort of shock effect which tank forces are designed to produce. However, defenders will often be forced to use tanks in small numbers in localised battles, in which case they should be integrated into the defence of a strongpoint or a series of strongpoints and given local protection by the infantry.

Artillery is likely to play an important part in the perimeter defence of a city; thus Forward Observation Officers (FOO) should remain on the perimeter or in high buildings so as to be able to control indirect fire. However, because of the intimacy of the urban battle, once the battle has moved further into the city and also because of the difficulties of observation, the employment of indirect fire becomes more difficult for the defender in the strongpoint battle. It is also worth mentioning that the Israelis in Beirut found that using artillery in the direct fire role was highly effective.

Engineers are particularly useful in urban areas, since they have the equipment and the expertise to prepare an effective obstacle plan. Although it does not necessarily require an engineer to construct many of the obstacle types needed for the successful defence of a town, nonetheless he is trained to lay mines and demolish buildings, erect wire, construct booby traps, blow craters, block sewers, clear routes and provide water. Engineers are invaluable in the planning of the defence.

Command, Control and Communications

Command, control and communications are unusually difficult in an urban area. Although movement between defensive localities and strong-points may well be difficult and dangerous, command must be close, personal and aggressive. Because of the likely difficulty of communicating once battle is joined, there is probably an even greater need for delegation of responsibility. Thus commanders will probably have to be prepared to act without recourse to higher authority for long periods. The techniques of mission analysis and directive control inevitably apply in urban warfare. Control in defence should be maintained by using simple plans, defined sectors and clear boundaries; this will, to some extent, overcome the fact that communications in an urban environment are likely to be inadequate. VHF radio will often be heavily screened and subject to highly reduced ranges. The use of high power radio sets, including HF, in armoured vehicles can offer some advantages over manpack sets. Quite often, though, it will be necessary to revert to either local telephones or line.

Fighting from Villages

An important aspect of defensive urban warfare is

the technique of fighting from villages. There is probably nothing that is intrinsically original about this technique, but it seems to have been developed into something approaching a theory in the 1st British Corps area of Germany in the mid-1980s, when the special characteristics of the Hanover Plain were adapted to create the 'pin table' metaphor. This describes an area in which villages are typically two to three kilometres apart, have perimeters of about 2,000 metres and provide good cover and concealment. Fighting from a carefully selected and prepared village of this type, infantry can deny easy access through it and destroy enemy tanks from the flank and rear as they move across the open ground between one village and the next. Thus enemy armour is 'bounced' from one village to another, much in the fashion of a ball on a 'pin table'. The idea was that the defence of certain individual villages (the pins on the table) should form part of a wider plan incorporating minefields, indirect fire, mutual support by long-range anti-tank weapons, and the use of armoured reserves. The overall coordination of such a battle would be at brigade or divisional level, with the battlegroups being given a cluster of villages to defend. These battlegroups would prevent the enemy from entering the vil-lages—instead the enemy would be canalised into suitable areas where they would be destroyed by mobile reserves, using the pivots provided by the villages as the base for their manoeuvre.

The defender undoubtedly has the advantage in urban combat. The close proximity of buildings limits observation which may be further reduced by smoke and dust. The defender who has chosen his ground and may have fixed fields of fire will have the advantage over the attacker who will be blundering around in a strange and complex environment. The defender who can take advantage of the opportunities for concealment will be difficult to locate, and it will be even more difficult to estimate his strength. On the other hand, the vertical dimension of the roof tops, cellars and sewers allows an attacker to bypass either over or under the enemy. Nevertheless on balance, a defender should have the advantage. Little has changed since the Rifle Brigade held the Wehrmacht at bay in Calais in 1940. They lost because they were grossly outnumbered. The force multiplying effect of an urban environment has its limitations. But in general, a defender outnumbered by three or four to one should be able to hold a city indefinitely. Stalingrad is a classic example of this truth.

8 The Employment of Artillery in the Urban Battle

The efficient employment of artillery in urban combat can be a battle winning factor. A classic historical example of the proper use of artillery in an urban situation was at Stalingrad.

The Red Army commander at Stalingrad made three key observations and based Soviet fire support countermeasures on these. First, the Germans were extremely uncomfortable with night operations, especially offensive exercises. Second, the German lines were vulnerable to patrols seeking intelligence, because they secured their positions poorly during periods of limited visibility. Finally, the Germans religiously followed their attack pattern.

Given these observations, the Soviet fire

British artillery firing in an urban battle in Italy during World War II

support system provided several countermeasures. During the night the Soviets became experts at infiltrating forward observers through the German lines. The Germans compromised most of their attack troop concentrations because of their poor noise and light discipline, and these concentrations became targets for Red Army artillery. Although there was little that the Soviet artillery could do to counter German airpower, it soon became apparent that if German routine could be interrupted it would often result in an attack being postponed. Thus, after the Luftwaffe had completed its morning air strikes, the Soviet artillery would fire a massive counterstrike at the troop concentrations, intelligence on which had been acquired by Soviet observation officers who had infiltrated German lines during the course of the night. These bombardments stalled the advance of German infantry and effectively suspended operations. The Katyuski multiple rocket launchers (MRLs) performed this counterstrike role very capably.

The Soviet artillerymen also learned to select the proper targets. Experience taught them that indirect fire had little effect against German tanks, so they concentrated on finding soft targets. The Stalingrad defenders quickly learned that destroying infantry, not tanks, stopped German attacks.

Soviet artillery commanders based their operations around artillery-strong points within the city, and these were chosen with several criteria in mind. First, they must dominate several hundred metres of street including, if possible, large intersections. Second, they must provide good observation and fields of fire for both direct and indirect fire weapons. Third, the artillery must be integrated with the infantry, armour and engineers in the strongpoint. And last, the guns must be sited so as to take advantage of all natural obstacles, primarily buildings and rubble, in the area.

The first countermeasure the Soviet fire support system employed was the extensive use of patrols and observation posts to detect targets. They placed these OPs to observe all avenues of approach; OP parties remained concealed within the rubble of the buildings to maintain observation, and the Soviet observers became masters of the art of camouflage.

The Soviets also used the buildings to conceal weapons from observation by German aircraft. Further, by using both buildings and rubble, the Soviets hardened their artillery positions against air and artillery attack. Buildings were especially important for protecting softer targets such as command posts, ammunition dumps and supporting units.

As soon as a German airstrike was over it was assumed that a ground attack would follow. Immediately, Soviet artillery would place fire on all known or suspected German observation posts, or artillery and armoured concentrations, and on advancing infantry in order to separate it from its supporting air cover. Soviet artillery would also fire automatically in support of any obstacles such as minefields or collapsed buildings. Finally, a proportion of the guns were readied to fire in the direct fire role if German tanks closed in on the strongpoint.

When the Soviets went over to the offensive at Stalingrad, their use of artillery was equally innovative. In order to support a surprise attack they would initially place fire on the objective to neutralise resistance, and would then shift fire in order to seal the objective from reinforcements. Once this was achieved, Soviet manoeuvre forces attacked the objective itself.

Soviet close artillery support consisted of both indirect and direct fire. As the maneouvre forces advanced, they were accompanied by direct fire artillery assault units. These ranged in size from a single gun to a battalion, and were controlled by the manoeuvre commander for use against bunkers, gun placements and tanks. This integration of artillery with the manoeuvre forces is foreign to western armies and is a particularly Soviet tactic. It is a primitive use of artillery, but it can be very effective—even in contemporary warfare, as the Israelis demonstrated in Beirut in 1987.

At times, an enemy position or obstacle *must* be reduced, almost at any cost. In such a situation, artillery weapons may be employed in the direct fire role: for example, in the absence of a suitable engineer vehicle or due to the relative ineffectiveness of tank ammunition against certain structures, one or two howitzer sections can be detached from the battery and join a manoeuvre element on a temporary basis. High explosive pro-

jectiles are usually the most effective ammunition in this situation. The 155mm projectile can penetrate up to 38in of reinforced concrete in a direct fire role, whilst an 8in howitzer can penetrate 56in of solid concrete. Special concrete piercing fuses are also required.

There are a wealth of lessons for modern artillery in urban warfare that may be drawn from the Soviet experience at Stalingrad. The gun section must develop expertise in close combat skills. Each section should learn how to achieve a high percentage of first-round, direct-fire hits against moving targets at very short engagement ranges. The gun section also needs to train as a rifle squad. On many occasions, gun sections at Stalingrad had to deploy as patrols, infantry squads and tank-

The effect of artillery fire and aerial bombing on Cassino in 1943. It was always debatable whether such destruction helped or hindered the attacking infantry

killer teams to respond to German penetrations.

The battery leadership must serve as both fire support element and strongpoint command. As a result, the commander must develop the skills of the artilleryman, infantryman and engineer. He must choose gun positions that allow his weapons to perform both direct and indirect fire missions. He can also take advantage of the urban terrain to enhance the survivability of his guns. Basements, warehouses and barns can house fire direction centres, supporting elements, vehicles and ammunition. The urban environment will help to prevent the enemy's rapid approach to a strongpoint but should one become untenable, withdrawal routes must be covered by fire from other strongpoints. Stalingrad also showed that pipes and culverts can provide protection for the strongpoints' wire communications. In a contemporary scenario, artillerymen must be prepared to use civilian telephone lines for backup communications, for as long as the system remains operational.

There are certain advantages and disadvantages for artillery locating in a built-up area. On many occasions, of course, an artillery commander may not have the choice—nevertheless, he should be aware of the factors which would influence his decision, some of them specific to artillery, some of them not.

On the credit side, a built-up area provides fortified positions and reduces the effect of enemy counterbattery fire. It also provides an obstacle to enemy armour. Most important, it degrades his detection capability—tracked vehicles leave little or no signature on hard surface roads, and enemy sideways-looking radar and infra red detection systems will have great difficulty in discerning between civilian and military signatures. Buildings provide overhead protection not only from the elements, but from view, from fire and—in a chemical warfare environment—from thickened persistent chemical agents. There are also particular advantages to locating guns in buildings with large double doors such as warehouses, factories, barns, garages and rail locomotive sheds. If these doors are oriented to or are near the battalion azimuth of fire, those buildings can be used as concealed firing positions. If not, they still provide excellent hide positions. A further advantage of urban terrain is that established road networks allow rapid occupation of gun lines and efficient resupply of ammunition.

There are a number of disadvantages, however, in locating in a built-up area. In Europe particularly, towns are located along the most obvious avenues of approach. In other words artillery units, whose job it is to keep out of direct contact with the enemy if possible in order to support manoeuvre forces more effectively, could be placing themselves in harm's way unnecessarily by selecting an urban location in which to deploy. Second, rubble may hamper movement within a built-up area: towed artillery is especially susceptible to restrictions caused by rubble. Self-propelled artillery cannot employ their spades (the device at the rear of the gun which is used to anchor a large artillery piece in place) on concrete or tarmac. Moreover, the orienting of artillery pieces can be impaired by the effects of electrical and telephone lines on compasses and aiming devices. Similarly, radio signals can be severely attenuated by buildings. And finally, a built-up area can be bypassed or isolated by the enemy, though arguably this might affect artillery less than direct fire weapons, since the guns can still be used to influence the battle 20–30km away irrespective of the local situation.

Artillery will never be the primary arm in the urban battle. But, if employed sensibly, and if due account is taken of its characteristics, artillery can influence the urban battle decisively.

9 The Employment of Armour in the Urban Battle

It has been estimated that 40 per cent of fighting in World War II took place in urban areas. Owing to the rapid urbanisation of Europe since 1945, it is calculated that in contemporary warfare 60 per cent of combat would take place in an urban environment. This sort of figure would hold good for most developed parts of the world, though clearly it would be much less in many parts of the so-called Third World. Thus it is neither logical nor sensible to emphasise the military truism that tanks are unsuitable for fighting in built-up areas. They may be, but the real point is that statistically they are bound to be used in the urban environment—quite simply because many wars are likely to continue to occur in towns and built-up areas, and because most armies have large numbers of

A British light tank, the Scorpion, mounting a 76mm gun while exercising in an urban training area known as Imber Village on Salisbury Plain Training Area

tanks in their inventories. The real problem is
how armour can best be employed in an urban
environment given its inherent unsuitability for
such a form of warfare.

Tanks

The tank has three main characteristics: protection, firepower and mobility, though different
nations have traditionally put a different emphasis
on these three characteristics. Thus the Germans
and Soviets have put mobility first, firepower second and protection third; they would argue that
extra mobility compensates for less armoured protection because a fast-moving tank is harder to hit.
Thus German engineers have designed tanks with
an impressive bhp, a good power-to-weight ratio
and consequently a good top speed moving across
country—the Leopard 1 and Leopard 2 have been
the result. The British, on the other hand, have
traditionally put protection first, firepower second and mobility third; this has arisen from the
British tendency to use tanks in penny packets to
'beef up' defensive positions, rather than in the
'Blitzkrieg' manner favoured by the Germans. The
Soviets have always shared the German philosophy; hence every tank since the T-34—the T-54,
the T-62, the 72 and the T-80—has been relatively small, fast-moving and agile, yet still man-
aging to deliver impressive firepower. Thus the
German Leopard and Soviet tanks have usually
weighed under 50 tons and been capable of speeds
across country of up to 50mph. The Americans
have tended towards the British set of priorities;
thus the M-60, A1 and A3 and now the Abrams
M1 and M1-A1 are tanks of some 60 tons and,
in all essential respects, the same animal as the
Centurion, Chieftain and Challenger 1 and Challenger 2.

In an urban environment mobility and speed
are not so important, and it is the tank's ability to
withstand punishment that is the crucial factor.
Certainly the German/Soviet tank design philosophy may well have been the best answer during
the forty years of confrontation between NATO
and the Warsaw Pact, since if a war *had* been

Soviet tanks advance through urban devastation on
the road to Berlin in 1945. The Red Army used tanks
en masse in urban combat

fought tank battles would have taken place primarily in rolling German countryside. However, now that any likelihood of confrontation between the two alliances in Europe has virtually disappeared, it is somewhat ironical that the much criticised British/American design in tanks may in fact be more suitable for some of the more limited confrontations which are likely to take place in the future.

The Israelis, for instance, favoured the British-designed Centurion during the 1973 Arab–Israeli war, because it was able to take a great deal of punishment—many hits on the Centurion just did not penetrate the armour. If a shaped charge projectile did penetrate the turret, it would usually result in the death or serious injury of the crew. But, remarkably, it was often possible to go on fighting a Centurion with holes in the turret and a new crew. They are exceptionally tough tanks, and in an urban situation this is just what is required. A tank may have to withstand many hits from hand-held infantry anti-tank weapons, and the front armour of a Challenger 2 tank can do just this.

For all their armour, however, tanks are vulnerable in built-up areas. They were designed to achieve shock and surprise in open countryside; when obliged to operate in an urban area they are most effective if they work closely with infantry, as the infantry can provide local protection for the tank—they can be its eyes and ears. Whilst communication between the infantry commander and the tank commander is perfectly possible on the radio, it is much more efficient and indeed reliable via his tank telephone. Most tanks—certainly all Western-designed tanks—have a small box on the rear which, if opened, reveals a standard telephone handset. This provides direct communication with the tank commander. An infantryman, with his vision unimpeded, can thus direct a tank commander, who may have his turret closed down and who is therefore dependent on vision blocks for acquiring targets, and guide him onto a target. This is called 'infantry-tank cooperation', and there are standard procedures for target indication. The usual method of getting a tank onto a target is to use the direction in which the tank gun is pointing as the initial reference, thus: 'reference tank gun, half-right, clock tower, bottom left-hand corner, enemy machine-gun'.

As well as dealing with targets which are proving troublesome for the infantry, tanks can be used for demolition or clearance purposes. Some tanks are equipped with a dozer blade and can be used to clear streets or rubble to improve mobility for wheeled vehicles. Alternatively if it is necessary to create a defensive position by blocking a street with rubble, tanks can use their dozer blades to erect a barricade.

Tanks have been widely used by the Soviets for internal security purposes, both in the Soviet Union and abroad such as in Afghanistan, Hungary and Czechoslovakia, though on all these occasions the effect upon public opinion of tanks or other armoured, tracked vehicles violating the streets was disastrous. But such considerations have not been of the highest priority for the Soviets. In the West, however, a different set of priorities applies.

Other Armoured Vehicles
Today, most armoured personnel carriers are tracked in order to perform their primary cross-country role in general war as efficiently as possible. However, tracked vehicles are not best suited to the internal security (or IS) role for a number of reasons. They are often difficult and expensive to operate and maintain; they are more noisy than wheeled vehicles; and they can cause damage to road surfaces. Most important of all, they are classed as 'tanks' by the layman and the media, and use of 'tanks' in an IS situation is often politically unacceptable; in most IS situations, vehicles are required to operate mainly on roads and in an urban environment, so that wheels are more suitable in every respect.

Most of these are 4×4 wheeled vehicles affording protection from small-arms fire up to and including 7.62mm, though some of the heavier IS vehicles afford protection against 7.62mm armour-piercing attack. They must all be provided with observation blocks so that the crew can see what is happening around them. In a conventional rural environment a vehicle is likely to be operating in wide open spaces in conjunction with many other vehicles and infantry on the ground, so it is not so vital that a crew has a comprehensive all-round view of the ground. How-

ever, in an urban environment an ideal IS vehicle *must* have good all-round vision—otherwise a petrol bomber, for instance, could very easily approach via a blind spot. Similarly, firing ports should be provided so that the crew can, if required, use their small arms from inside.

Vulnerable points such as the fuel tank and the radiator should be given special protection, particularly from petrol bomb attack. The other main threat is from anti-tank grenades. Certainly in

A Short's Shoreland IS vehicle is seen here on airfield guard duties. The Shoreland has been used by the Royal Ulster Constabulary in urban riot situations

Northern Ireland, the IRA have used RPG-7 rocket launchers against APCs, which are insufficiently armoured to prevent penetration by projectiles from rocket launchers. However, in an urban environment the close range at which terrorists are forced to engage APCs militates against

Talon security vehicle with water cannon, grenade
launchers, arclights, anti-riot grilles and barricade
remover

a successful engagement. The limited exposure time of an armoured vehicle passing a fixed point means that the firer has very little time to recognise the target, prepare to fire, acquire the target, aim and engage. Often RPG-7 projectiles have passed behind their target, on some occasions unnoticed by the occupants of the vehicle.

An IS vehicle must be so designed as to enable crew and passengers to get in and out quickly, and there are many examples of models where this characteristic has been included in the design. Clearly, in a confused situation probably involving large and disorderly crowds, it is only sensible to have multiple doors. In a conventional war situation, the enemy is normally expected from a single direction; in a guerilla situation, the enemy may attack from any quarter and the requirement is therefore to be able to leave the vehicle from the opposite side to the direction of the attack. Side doors dictate that there should be only four wheels, an arrangement that is also sensible in the interests of simplicity and mechanical reliability. IS armoured vehicles can be fitted with a variety of armament installations, including water cannon, tear gas launchers, and machine-guns. Some can even be electrified to prevent rioters climbing on to the vehicle.

In addition to armoured vehicles, there are several other types of vehicle commonly used in IS situations: water cannon vehicles, which may or may not be armoured; conventional 'soft skin' vehicles that have been covered in a form of appliqué lightweight armour as protection against blast and low-velocity rounds; and armoured bulldozers for the removal of barricades. Appliqué armour was first developed for the British Army in an attempt to provide some protection for Land Rover crews against blast, fire and acid bombs, and low-velocity small arms fire. For example, GRP is a form of fibreglass used to cover the body and roof of Land Rovers, while Macralon, a form of strengthened plastic, is used to cover windscreen and windows.

Older vehicles can be adapted in many ways for IS purposes. In Northern Ireland, for example, the British Army has adapted the long-serving GKN Sankey AT-104 (commonly known as the 'Pig') by attaching unfolding fenders to each side of the vehicle. If it is parked in the middle of a relatively narrow road flanked by buildings, it can successfully block off most of the road, and can also afford protection against missiles thrown by rioters. The adapted vehicle is known as the 'Flying Pig'. Other possible attachments are roof- or turret-mounted searchlights, loudspeaker systems and a strong device for removing barricades. A self-help device that is fitted to many jeeps and Land Rovers in the IS role is a fence picquet—this is attached vertically to the front of the vehicle to cut steel wires that have been stretched across roads, a hazard intended to cause serious injury to the occupants of open vehicles.

A common threat is the land mine. The design of IS vehicle hulls should ideally be such that, if a mine is triggered off by one of the wheels, the upward slope of the hull should deflect much of the blast, and the strong monocoque structure should provide maximum protection so long as the crew are strapped into their seats. Examples of a shaped hull are the South African Hippo vehicle, the British GKN Sankey Saxon, and Italy's Fiat 6614CM APC.

IS vehicles are normally of simple and rugged construction; they are therefore often employed by less developed countries with limited maintenance resources. A vehicle that has followed such a design philosophy closely is the GKN Sankey Saxon. It is powered by the widely available Bedford 500 6-cylinder diesel truck engine, and because it is constructed of commercially available automotive parts, it means that anyone who can maintain a truck can also maintain the Saxon. Similarly, the French Berliet VXB anti-riot vehicle uses 'off-the-shelf' Berliet truck spare parts and is both easy and cheap to maintain. Design details are very important. For example, in the Belgian Beherman Demoen BDX, the engine air intake is located below the generous canopy over the driving position and has a moving shutter to provide further protection against Molotov cocktails. The twin exhaust pipes run along the two sides of the roof to make it more difficult to climb on to the vehicle.

Increasingly, anti-terrorist forces throughout the world are recognising that 'discreet operational vehicles' (DOVs)—standard commercial vehicles and limousines that are armoured without appearing to be so—are less provocative for

The AT-104 operating in Belfast in the mid-1970s

urban internal security situations than the more heavily armoured, obviously military hybrid vehicles. Certainly there will continue to be IS situations that warrant the attendance of highly protected military vehicles, but in many incidents the use of DOVs would be more politic and just as effective. Examples might include confrontation with lightly armed terrorists, student demonstrations, and the carriage and escorting of government VIPs. The current range of DOVs includes:

1   Land Rover- and Range Rover-type vehicles for anti-terrorist operations, with a cross-country capability and all-round protection against handguns, submachine-guns, grenade fragments and certain categories of rifle.

2   VIP limousines with all-round protection against hand-guns, submachine-guns and grenade fragments.

3   Saloon cars and Range Rover-type vehicles for VIP escort duties with partial protection against hand-guns, submachine-guns and grenade fragments.

4   Fast patrol cars for immediate response and pursuit, with frontal protection only against handguns and submachine-guns.

(Opposite, above and below) A British Chieftain Main Battle tank of the Berlin Brigade in urban camouflage. Parked in the middle of the square it stands out, but in an urban landscape the rectangular camouflage pattern is surprisingly effective

Modern technology is narrowing the gap between the technically feasible and the operationally desirable in terms of both opaque and transparent armour. However, thought does need to be given to the design of a DOV: the answer is not necessarily to cram as much armour as possible on to a given chassis. Users often tend to ask for unrealistic and often unnecessarily high levels of armour, while insisting on minimum changes in vehicle performance and appearance—in reality the two are often irreconcilable. On the other hand manufacturers, whose experience is usually confined to automotive engineering, tend to offer a solution that does not take sufficiently into account the operating conditions and protection requirements of the user. If, for instance, a manufacturer decides that the main threat to a head of state is from the 7.62mm NATO rifle or its equivalent, the passenger section of his armoured limousine can be given complete protection against single shots from this weapon. But in order to avoid significant modifications to the engine and the suspension, the driver's section would need to be left unprotected. An IS expert would advise that a driver killed or incapacitated when driving at speed would probably result in the death of the head of state anyway! Moreover, where the protected limousine is for a head of state, the need for protection to this degree can be questioned: sensible precautions along a route should reduce the opportunities to use a high-powered rifle, and where such use is very likely, the VIP should change his route or cancel the engagement altogether. A much more likely threat is from the assassin in the crowd armed with a concealed hand-gun or submachine-gun; the wiser bet would therefore be all-round protection against *this*, rather than armouring only the rear section of a limousine against the unlikely use of a 7.62mm weapon.

Successful DOV design should start from the premise that no DOV will be bulletproof, and that the best protection will merely buy time. The armoured Lincoln Continental limousine delivered to the US Secret Service in 1969 carried two tons of armour steel and bullet-resistant glass, and was capable of travelling at 50mph with all tyres shot out. However, the Secret Service would have been the first to admit that its most attractive characteristic was its ability to maintain mobility with all its tyres deflated—the mass of armour could not in fact have protected the occupants from sustained fire from high-power automatic rifles available on the open market. It might have resisted the first few rounds, so giving time for the agents travelling in the back-up vehicles to return fire, or for the chauffeur in the protected limousine to put his foot on the accelerator.

Some of the problems that arise in attempting to bring about a compromise between discretion, protection and performance will be apparent from the table on p.117. This gives the ballistic properties of a selection of weapons used by terrorists, and the necessary thickness and weight of various armoured materials currently in use which would be capable of resisting them. Lighter materials with similar or superior resistance qualities are under development, but they are not yet widely in use. It can be seen from the table that even all-round protection against a low-velocity 9mm SMG will impose a considerable weight penalty. Such protection would seem to be a sensible minimum for VIP limousines, with the possible addition of selected points being protected against 7.62mm high velocity rifle attack in high-risk areas of the world. It is possible to compensate for the increased weight by restricting the number of persons in the vehicle or by modifying the chassis, suspension and engine. If the occupants of a DOV are in a position to return terrorist fire then there are different options.

A fast patrol car may only need a bullet-resistant windscreen, an armoured engine bulkhead, and armour added to the inside of the front doors and the rear of the front seats. Providing only two persons use the car, it would normally require no modification to engine or chassis in order to achieve the same performance as the original version of the car with four persons in it. An escort vehicle, however, would normally require more extensive protection, if only because it might have to serve as a temporary refuge for a threatened VIP.

Wheels are particularly vulnerable points on a DOV. Some sort of 'run-flat' capability is essential for accidental or induced blow-outs at speed and to enable the driver to extricate the vehicle from

an ambush if the tyres are shot out. There are various solutions: one involves the fitting of steel discs inside the tyres so that the weight of the vehicle is supported when the tyres are deflated; another is the Dunlop Denovo system, which injects a lubricant between the deflating tyre and the wheel rim. Both systems allow the vehicle to be driven out of an ambush—the Denovo system permits it to be driven up to a hundred miles.

There is no doubt that DOVs have their place in the fight against terrorism in an urban situation, particularly as assassination or kidnapping of VIPs is now a favourite terrorist tactic. Ostentatious personal protection precautions can be counterproductive in PR terms, particularly for a politician, and are likely to be exploited for propaganda

One of the few occasions on which main battle tanks have been used by the British Army in an IS situation was during operation Motorman, which took place on 31 July 1972 in Londonderry. Its aim was to destroy the 'No Go' areas which had been set up by the IRA in that city. Centurion Engineer vehicles, mounting a bulldozer but no guns, were used to demolish barricades

purposes by political dissidents. DOVs, on the other hand, can provide an effective low-profile alternative.

In conventional warfare, therefore, tanks perform a useful function. In an urban situation, internal security (IS) armoured vehicles—whether wheeled armoured cars, armoured Land Rovers or DOVs—fulfil an indispensable role. They provide protection for the security forces; they are command and control vehicles; they mount anti-riot weapons; they are used to clear streets; they act as armoured ambulances; they may even be used to block streets. There is little doubt that armour will continue to play a vital role in urban warfare.

Terrorist weapons and armour materials used to resist them

Weapon	Muzzle Velocity	Impact Energy at 50m	Aluminium		Steel		Composite		Glass		Glass/ Polycarbonate
	m/sec	mkg	mm	kg/m²	mm	kg/m²	mm	kg/m²	mm	kg/m²	mm
Pistols											
9mm Luger	338	47	4	11	2	14.58	6.86	13.7	25	61	–
.38 Colt	260	35	6	16.6	2	14.58	6.86	13.7	25	61	–
.38 S & W	185	16	4.5	12.5	2	14.58	6.86	13.7	25	61	–
.357 Magnum	439	101	7	19.5	2	14.58	8.66	17.1	25	61	–
Submachine-guns											
9mm Sterling	390	64	7.5	21	2	14.58	8.66	17.1	29	70	–
.45 Thomson	280	58	7.5	21	2	14.58	8.66	17.1	29	70	–
Rifles											
7.62 Ball Nato	855	380	27.5	70	5	39.38	15.23	42.3	63	150.7	35.5'
5.56mm Armalite	990	173	21	59					63	150.7	35.5'

Notes: A curved windscreen of glass-polycarbonate mix now available is capable of stopping single shots from the NATO 7.62mm rifle and the 5.56mm Armalite (both ball). The armour data is drawn from international sources. Resistance figures are derived from tests held under varying conditions. BSS 5051 Part 1 (1973) details appropriate performance requirements and test methods for security glazing. It makes provision for spall from the rear surface when attack takes place. These stringent conditions are not universally applied.

10   Techniques and Tactics of Urban Counterterrorism and Riot Control

Cities provide a focus for protest. Urban terrorism is the most extreme form of protest and urban rioting a lesser form, although one can often lead to the other. Rubber bullets and tear gas are used to contain urban rioting, bomb disposal teams operate against the urban bomber, and Special Warfare or SWAT Teams operate against urban terrorists or armed criminals: all are examined in this chapter.

Rubber Bullets

The 'rubber bullet' first came to the public's attention in the early 1970s when it was introduced into Northern Ireland for use by the security forces to control riot situations. In fact, although rubber bullets had not been widely used, some police forces in the United States were already equipped with similar systems. Nevertheless, the first wide and sustained use *was* in Northern Ireland.

In a riot situation it is usually preferable for troops or police to maintain a reasonable distance between themselves and the crowd. This prevents them being overwhelmed or outflanked, besides which tempers tend to remain cooler if a sensible distance between the two sides is maintained. This is not always possible, and when close contact is unavoidable, the most common means of crowd control has been the wooden baton or truncheon. However, the close physical proximity concomitant with the use of a hand-held instrument in fact inflames passions, allows insults to be exchanged, and often results in situations getting out of hand. One way around this problem is to use grenade launchers to project CS gas grenades into a crowd. In extreme situations CS gas is a highly effective crowd dispersant, but it affects

Two versions of the baton round or rubber bullet are shown top left in this photograph. Both the single round and the multi-baton round are shown

bystanders, adults and children, rioters and (if they are not equipped with gas masks) security forces alike. Moreover, crowds all over the world have become adept at wearing wet handkerchiefs over their faces to combat the effects of the gas, and have learned to throw or kick the grenades back at the security forces. The advantage of the rubber bullet, on the other hand, is that it is selective. It is accurate up to a range of about 90 metres and is designed to cause no more than bruising or shock.

When the British Army was first deployed in Northern Ireland in 1969 in support of the civil power, army manuals still outlined crowd control methods that had been appropriate for imperial policing duties. In a riot situation a long way from home it had been traditional to warn the crowd to disperse over a loudhailer, using banners pro-

claiming 'Disperse or we fire'. If, in due course of time, the crowd did not disperse, the ringleader or ringleaders were shot—not surprisingly, this usually did the trick. But such methods were hardly acceptable within the confines of the United Kingdom where the news media could instantly relay a developing situation onto TV screens in millions of homes. Double standards maybe, but a fact of life. Rubber bullets therefore offered a way out.

What exactly are rubber bullets? Originally they were developed to deter individual petrol bombers or stone-throwing rioters at ranges of up to 60 metres, and at first came as two distinct types: the rubber baton round and the plastic baton round. The former was intended to be a cheaper alternative to the latter, and consisted of a hardened rubber cylindrical projectile sealed in an aluminium cartridge to make it waterproof. The calibre was 38mm, projectile length about 10cm and weight 175g. Early anti-riot guns, such as the Schermuly 1.5in, projected these rounds at a muzzle velocity of 100mm/sec. However, the plastic or PVC

A rubber bullet inside and outside its cartridge case. These are designed to 'discourage' rioters and should only be fired at individuals over 30 metres away if serious injury is not to be caused

baton round has now almost entirely superseded the rubber round. Its dimensions and weight are essentially the same, but it is more accurate. However, the more generally accepted term 'rubber bullet' will be used here.

The rubber bullet is an extremely simple item of equipment. Much more interesting and variable are the means of delivery—grenade launchers—that are on the market. There are many, manufactured for the most part in the United Kingdom, all of which are designed to project a rubber bullet as accurately as possible over a distance of about 60–90 metres. A brief review of four systems which are representative of the broad range of weapons available is the best way of focusing on grenade launchers.

The first is the **Arwen Ace**. This is a single shot anti-riot weapon manufactured by Royal Ordnance which fires the complete range of purpose-designed Arwen ammunition: AR1, which is a simple rubber bullet; AR2 which lays down a carpet of irritant smoke; AR3 which combines a body blow with a discreet dose of irritant; AR4 which dispenses screening smoke; and the AR5 which is a barricade penetrator. Loading is achieved through a single aperture, eliminating the need to 'break' the weapon to load it. It is a 37mm weapon and can fire twelve rounds per minute, and its maximum range is 100 metres. The latest version of the Arwen is the Arwen 37, essentially the same weapon but with a five-shot magazine added, thus eliminating the need to reload after each shot. The Arwen is in use with the British Army in Northern Ireland, and has also been exported to a number of other countries.

A second example of grenade launcher is the **CIS 40 GL**. This is a 40mm weapon which, like the Arwen, can fire a wide range of grenade cartridges out to a range of about 380 metres, though the rubber bullet round would be limited to about 100 metres. What makes the CIS 40 GL interesting is that it is manufactured by Chartered Industries of Singapore, and is therefore another example of the increasing tendency of so-called Third World States to meet their own weapons requirements. This of course has further implications for those nations which traditionally produce defence equipment, such as the United States, the Soviet Union, France and Britain.

Another method altogether of projecting rubber bullets is to launch them from a conventional rifle. Israeli Military Industries have developed a simple but highly effective system: the rubber ammunition—in this case not bullets as such, but 15 cylindrical plugs—is packed in an aluminium launching container. This in turn is fitted with an adapter tube which slips easily over the flash suppressor on a rifle. The rubber projectiles are ejected from the launching container by the gases from the ballistic launching cartridge. After firing, the launching container remains on the rifle and can be easily removed by hand. The cloud of projectiles can be used over distances of up to 90 metres, and can be fired from most rifles including the M-16, Galil and FN. The weapon differs from the standard rubber bullet in that it is an 'area weapon', that is to say it is fired at the crowd, not at an individual. Like the rubber bullet, though, it is designed to scatter the crowd, and causes some pain to those whom it hits.

The **Excalibur multi-shot riot gun** is a 37/38mm weapon employing the well proven revolver action. It is a simple and rugged weapon manufactured by Wallop Industries and capable of a high rate of fire. It allows the firer to engage several targets, and is more effective than the standard single shot break-open riot gun.

There was one other attempt to develop a non-lethal projectile to deter demonstrators. The **ring airfoil grenade** (RAG) was developed in the mid-1970s at the Aberdeen Proving Ground in the United States as a means of controlling civil disturbances without close-up confrontation. RAG projectiles were fired from a launcher attached to a standard M-16 rifle used at that time by the US Army and National Guard, as well as by numerous state and municipal police departments. There were two versions of the projectile, known as 'soft' and 'sting'; both were developed from a thick, one-piece body of soft rubber material shaped like an aerofoil and rolled into a ring, and were developed to hit an individual at ranges varying from point blank to 50 metres, or to hit small groups at twice that distance, producing pain but unlikely to cause serious injury. Both projectiles weigh the same and have the same dimensions, and are launched spinning at 5,000 rpm. This provides gyroscopic stability during

flight. The soft RAG is identical to the basic sting version, except that it contains a small quantity of CS powder.

Whatever the type of rubber or plastic ammunition, it is important to understand the context in which it is designed to be used. It is intended to be a weapon which allows security forces to *de-escalate* a situation. Clearly high velocity bullets are designed to kill. Equally CS gas is highly provocative. Rubber bullets, on the other hand, provide the security forces with a means of hitting back at rioters hurling bricks and petrol bombs, and all the evidence to date is that rubber bullets are an effective means of dispersing crowds and *do* defuse situations.

There have, however, been a few cases where rubber bullets have caused serious injury or even death. In situations in which police are seriously outnumbered and surrounded, and are forced to fire their riot guns at too short a range—at less than 20 metres—serious injury can result. It depends on which part of the body the rubber bullet hits the victim and on a number of other factors such as, for instance, the state of health of the victim. In Northern Ireland there have been a very few deaths resulting from the use of rubber bullets, probably only two or three. If this is put in context—the thousands of rubber bullets fired during some months in Northern Ireland, resulting in no injury at all, and the twenty-two years of conflict—then the rubber bullet must be considered both an effective and a relatively safe projectile (insofar as any projectile can be considered to be safe). As long as it is accepted that the security forces must be given some means of keeping threatening crowds at bay, then the rubber bullet is probably the best solution.

A glance at a diary of events in Northern Ireland shows that a day seldom goes by without some rubber bullets being fired. For instance on the 9 August 1988, which was the 17th anniversary of the introduction of internment in Northern Ireland, rioting occurred in the north of Belfast in the New Barnsley area of police responsibility. There was rioting in the Ardoyne and New Lodge areas as well as in the Falls Road, the Markets and the Ballymurphy areas. Military

CIS 40GL grenade launcher

patrols fired 339 rubber bullets and the Royal Ulster Constabulary at least 238 during the night. Nevertheless there were no apparent casualties and no injuries were reported. This was an unusually active twenty-four hours; in most situations only a few rounds are likely to be fired. Whatever the situation, specific orders must be issued—certainly in the British Army—before rubber bullets may be used. They can be dangerous, and clearly their use must be carefully controlled. A rubber bullet that kills is a tactical failure.

Rubber bullets used responsibly and at the ranges for which they are designed are safe. They provide otherwise powerless soldiers and policemen with a means of controlling riots and, equally important, hitting back at troublemakers. In the vast majority of cases they solve and defuse situations without causing any injury. In a tiny minority of cases rioters have been badly hurt and, in a very few cases, fatally injured. However, it would be difficult to design a better system than the modern rubber bullet or, more precisely, baton round, and it has now almost entirely superseded the alternative RAG and rifle-launched grenade systems. As long as civil disorder remains a phenomenon of our times, so rubber bullets are likely to remain in the armoury of the forces of law and order.

Tear Gas

The irritating or harassing agents, which are more commonly described as vomiting or tear gases, are sensory irritants. Only lethal in extremely high concentrations, their action is usually rapid although their effects are comparatively brief in duration. Some agents cause a temporary flow of tears and are known as lachrymators, some induce sneezing and coughing (sternutators), some irritate the skin to cause severe itching or stinging sensations (orticants) and some, if they are swallowed, precipitate bouts of violent vomiting. Some of these agents were developed during World War I, notably chloroacetophenone (CN)—the 'classical' tear gas—and Adamsite (DM) which acts more slowly than CN but produces more severe effects including lachrymation, salivation, headaches, nausea and vomiting.

Even more effective than CN and quicker

The tear gas grenades illustrated here are standard versions, designed to be thrown by hand rather than projected from a launcher

acting than DM is Orthochlorobenzylidene Malonbuitvise, known as CS after its discoverers, B. B. Corson and R. W. Stoughton. Though first discovered in 1928, its present form stems from work conducted in Britain in the 1950s when the War Office requested a more effective, but preferably less toxic, riot control agent than CN. The effects of CS occur almost immediately and, depending on dosage, range from a prickling sensation in the eyes and nose to a gripping pain in the chest, an uncontrollable flow of tears, coughing, streaming nose, retching and vomiting. Few people are prepared to put up with the symptoms of CS for more than a minute or two if the concentration is higher than 2 milligrams per cubic metre, although after repeated exposures the

body can tolerate concentrations up to five times that figure. In normal circumstances, however, individuals are incapacitated within 20–60 seconds, and the effects last for a further 5–10 minutes after the affected individual is removed to fresh air.

The gas is used almost exclusively by the forces of law and order, whether police or military, in internal security situations; the only use to which it has been put in actual warfare is in winkling guerillas out of tunnel systems. The Americans used it for this purpose in Vietnam. Today it is used in four basic situations: its most common use is for riot control and mob dispersion—with proper training, a small, well-equipped unit of policemen can control and disperse almost any dangerous mob with tear gas. The second use is in a siege situation, when an armed criminal or deranged person has barricaded himself in a building and refuses to come out. The third is in the control of prison riots; although CS gas has not been used in this context in the United Kingdom, it has in many other countries, particularly the United States. Tear gas can also be used as an alternative to a hand-gun, in either an offensive or defensive situation, for instance to arrest a criminal or as a means of self-defence. It is in this context that Chemical Mace non-lethal weapons have been widely accepted throughout the world: as such, CS can be sprayed in the form of an aerosol from a hand-held pressurised container.

Tear gas is, of course, controversial. It is indiscriminate and affects rioters and innocent bystanders alike. For this reason it has not been widely used in the United Kingdom, except in Northern Ireland; and even there, rubber bullets are used in preference to tear gas because individuals can be singled out as targets. However, Britain appears to be almost alone in having these qualms about tear gas. In Europe, all police forces are equipped with it, and it is regularly used to disperse riots and illegal gatherings. Similarly, both the police and National Guard in the United States have used tear gas fairly regularly. It is also quite common for individual citizens, particularly females, to carry an aerosol container of Chemical

These two versions of the 'stun' grenade are used by special forces and SWAT teams to cause a diversion whilst entering a building or aircraft occupied by terrorists or armed criminals

Mace on their person for use against an assailant. This is entirely legal in the US; in the UK it is classified as an offensive weapon and is illegal. Tear gas is also standard equipment in the police forces of Central and Eastern Europe, and during the revolutions of 1989, many of the regional police forces used tear gas against protestors. It is also commonly used in South America, in Japan, in India, in Israel and in many other countries. Whilst some provincial police forces in England, Scotland and Wales are equipped with tear gas, it has not been used to date on the British mainland, although it has been deployed in various siege situations. Although it is certain that the SAS used stun grenades in the 1982 assault on the Iranian Embassy, it is not known for sure if CS gas was used. Since gas masks were worn by the SAS troopers in the assault, it is probable that CS gre-

Smith and Wesson 'pepper fog' tear smoke generator. Also illustrated are a Smith and Wesson gas mask, police helmet, 37mm gas gun and projectiles, and various gas grenades

nades were also used. Surprisingly, CS gas has not been used in any of the recent spate of British prison riots. In view of the length of some of the sieges and the degree of damage done, it is in retrospect quite extraordinary that CS gas was not used, as it would almost certainly have resulted in a quick end to the disturbances. Tear gas has, however, been used on several occasions in Northern Ireland, particularly in the 1970s, although more recently, rubber bullets have become the preferred method of crowd control in the Province.

There are various methods of dispensing tear gas; the most common is the grenade launcher or gas gun. This can be either a purpose-designed grenade launcher—of which there are many on the market—or it can be an add-on device to a standard service rifle which allows it to project a CS grenade. Grenade launchers are manufactured in many countries, including the US, UK, Israel, South Africa, France, Italy and Singapore. They are mostly multi-shot, 37mm calibre and can project a grenade up to about 150 metres. They

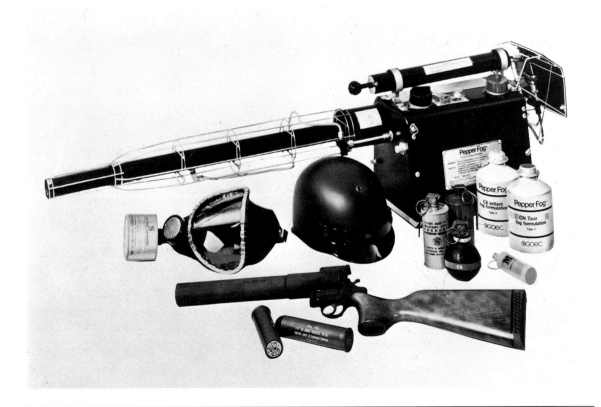

HG40 grenade launcher

project two categories of grenade, the standard cylindrical grenade and the rubber ball model; the former discharges CS or CN gas over a period of usually approximately 20 seconds. The latter does the same, but has the added advantage that it bounces all over the place and is almost impossible to pick up and throw back. All these grenades can, of course, be thrown by hand, though their range would then be limited to 20 yards or so; and this would mean closing with the crowd, which is not ideal.

The third category of dispensing tear gas is by means of the hand-held 'Fog Ejector'. Special CS fog and compressed gas is stored in a cylinder device which is hung from a shoulder belt and sprayed under pressure from a nozzle; obviously the operator must wear a gas mask. It contains special CS particles which are designed to float in air, has a range of about 15 metres, and is suitable for use in confined spaces and narrow alley ways. Fog ejectors are capable of up to 40 one second bursts before recharging. The Israelis use these systems widely in the narrow streets of Jerusalem and other Israeli towns.

The last method of dispensing tear gas is by means of a shot-gun cartridge. These are designed to be fired from all standard 12 bore riot guns. Although their capacity is clearly much less than that of a grenade, they dispense CS gas for up to 20 seconds and are immensely accurate. They can,

for instance, be fired through a two foot diameter circle at a range of 80 metres by most marksmen. This means they can be fired through a window in a siege situation. They can penetrate ¾in plywood at 50 feet, or window glass up to 80 metres.

The tactics of using tear gas are based on the overall need to maintain a distance between the security forces and crowd. Once these two become embroiled, tempers tend to flare, insults to be exchanged and heads get cracked. Whilst the effects of tear gas are unpleasant in the short term, serious injury is avoided, crowds are dispersed and ugly situations quickly solved. To be effective, tear gas is best fired in volleys so that the effect of the gas is overwhelming—just the odd canister of CS gas can be avoided, kicked away, or it simply fails to achieve a concentration that is sufficiently uncomfortable for enough people. Wind direction should also be taken into account, otherwise the crowd may be driven *into* the security forces, thus compounding rather than solving a situation. Snatch squads equipped with gas masks are best dispatched into a crowd to arrest ringleaders or troublemakers in the wake of a CS barrage when the crowd is disorientated and temporarily incapacitated.

A crowd's reaction to tear gas is dependent upon the extent to which its members have been

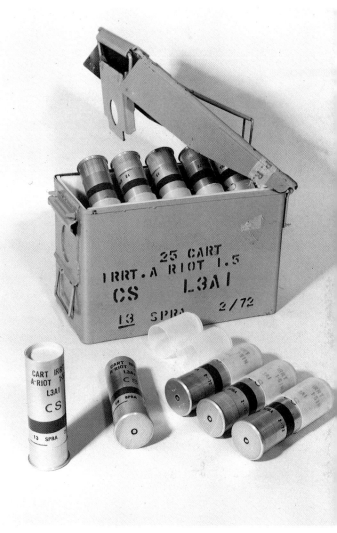

Schermuly 1.5in anti-riot irritant cartridge

A stone-throwing mob in Belfast attacks British troops deployed in the anti-riot role. The aim in this sort of situation is to maintain a distance between rioters and security forces. This can be achieved by projecting rubber bullets or CS gas from grenade launchers

exposed before. Some protestors in Japan and South Korea, for instance, are old campaigners when it comes to combating the effectiveness of CS gas. Individuals wear wet handkerchiefs around their noses and mouths which mitigate the effect of the gas somewhat. Rioters pick up grenades and throw them back at the security forces—some even possess their own gas masks. In Northern Ireland, youths became so used to exposure to CS gas that their tolerance has increased four- or five-fold. However, those that have never experienced the effects of tear gas before, not surprisingly become extremely agitated when they start suffering from the symptoms, many to the extent of extreme panic. This is particularly the case if children are affected. Those with asthma would do well to avoid situations in which they might get caught up in a crowd that is in danger of being tear-gassed. Although it is unlikely that an asthma sufferer would die from the effects of CS gas, his discomfort would be extreme, possibly to the extent of him losing consciousness. Certainly in Europe police are instructed to search for victims that may be suffering unduly as a result of a tear-gas barrage; it is doubtful, however, that such consideration for the individual is exercised in many Third World countries—victims there are left to crawl off to hospital unaided.

Tear gas should not be used except in serious riot situations. Arguably it is used too easily and too often in many Third World countries—this not only devalues the currency and makes it less of a deterrent in the short term, but also increases the risk of injury and the dangers of exacerbating already dangerous situations in the long term. However, very often the British authorities seem to go to the other extreme and not use it at all. There are many examples of prison riots in the UK where early and judicious use of CS gas might have saved lives, damage and injury. There is, as always, a middle course, and there are several situations in which tear gas is particularly suitable. Prison riots and siege situations are probably the best examples. And the use of tear gas, providing this is governed by the appropriate rules and safeguards, need not in any way indicate that a state lacks legitimacy: just as lethal force may have to be used ultimately in certain situations in order to

avoid entirely innocent or greater loss of life, so tear gas provides another rung on the escalation ladder which puts off the necessity of using lethal force either too early or at all. In our increasingly violent society, CS gas is a necessary evil.

Remote Control Bomb Disposal
One of the most widely used terrorist weapons is the bomb. Bombings are aimed at military targets (vehicles, soldiers or bases) or economic targets— or they can be indiscriminate. The rationale behind an attack on a military target is clear enough. Those against economic targets incorporate a longer-term strategy, and are designed to undermine the ability and determination of the ruling state to carry on the war against terrorism— clearly the more expensive a campaign becomes, the more difficult it is to justify to the taxpayer in a democratic state. For example, the IRA has regularly bombed factories, shops and other economic targets in Northern Ireland. Indiscriminate bombing is, perhaps, terrorism in its purest form; by creating indiscriminately an atmosphere of terror it is hoped to so intimidate a population that they will be cowed into submission. In recent years aircraft, shopping precincts, hotel lobbies and/or airport check-in facilities are but a few of the many public places that have been bombed by the Red Army Faction, the Bader-Meinhoff Group, Palestinian terrorists, Sikh extremists, the IRA and many other terrorist groups.

The type of terrorist explosive device varies according to the nature of the planned target and the skill of the bomber; it could incorporate commercial or home-made explosive materials and be initiated by command detonation, a timing device or target influence. Perhaps the method of transporting and placing a bomb which has become the most commonplace as a means of terrorist attack across the world is by using a car (the car-bomb). And in the Lebanon in 1983–4, even larger amounts of explosive were packed into lorries and driven by suicidal Palestinians straight at buildings occupied by US and French troops of the four power peacekeeping forces. On one particular occasion the method proved horrifyingly effective when 241 US marines were crushed to death in their collapsed barrack block. A variation of the car-bomb is the so-called 'proxy' bomb, a

A top view of an IRA nail bomb. The device is designed to scatter lethal 6in nails over a wide radius

technique developed by the IRA in Northern Ireland—the terrorists intimidate an individual (usually by holding his family hostage under threat of death) into driving a car-bomb up to a target and leaving it there. 'Proxy' bombs are normally activated by means of a timing mechanism.

The 'command wire improvised explosive device' (CWIED) and the 'radio-controlled improvised explosive device' (RCIED) are the most difficult types of bomb to guard against. The first requires the individual initiating the bomb to be at the end of a wire and to be able to see the target; the second provides the bomber with greater flexibility, but it also requires more sophisticated initiation equipment. Much research is going on into ways of countering the RCIED. The third main category of initiation is known as target influence, whereby the presence of the target initiates a trembler or similar device.

The home-made mortar is another method of delivering high explosive to a target, and one

A homemade IRA multiple mortar mounted on the back of a lorry. This is similar to the type used to attack 10 Downing Street in February 1991

which is much favoured by the IRA. Guerilla movements all over the world have used mortars to attack targets remotely, but have usually been able to acquire properly manufactured mortar systems, either from the international arms market or by acquiring them in action from government troops. In Northern Ireland it has to date proved too difficult for the IRA to smuggle such bulky weapons into the country. In any event, the sophisticated sighting systems of manufactured mortars make them unnecessarily complicated for the IRA's simple requirement—to lob high explosive a short distance into a security force base. Ingeniously the IRA has developed instead a series of home-made mortars, which although unreliable and unpredictable, have sometimes managed to inflict casualties inside security force bases. Their only use on the UK mainland was the attempt in February 1991 to lob a bomb onto 10 Downing Street. The bombs missed—but only just.

A fire bomb exploding in front of an IVECO FIAT SS-13 security vehicle

There is a distinction between a bomb designed to cause maximum damage to buildings or military vehicles, and one that is targeted against an individual. The former consists of hundreds of pounds of explosives and can be secreted in a vehicle or a large container. The second is altogether a smaller, and often more sophisticated device which can be attached to the underside of a car, to a door or even delivered by post. It is a common terrorist tactic to place bombs underneath cars or in the engine compartments of cars belonging to members of the security forces during the night, so that when they drive them away next morning a trembler device detonates the explosive. This method was used to murder Ian Gow, a Conservative member of Parliament, outside his house in the summer of 1990. There is a third category of bomb, too: the incendiary. This is an even smaller device with a small timer attached to it, which can easily be placed under inflammable materials; it bursts into flame for long enough to start a fire in the target building.

The skills of bomb detection and disposal have

developed rapidly since the early 1970s. The British Army, like other major armies of the world, already had bomb disposal experts before the emergency in Northern Ireland, mostly employed in disposing of World War II German bombs. Responsibility lay (and still does) with the Royal Engineers for dealing with unexploded bombs on the battlefield; the Royal Air Force deals with bombs on its own airfields, and the Royal Navy with mines in the sea and on the beaches. But the dangerous task of dealing with improvised explosive devices (IEDs) has always been the responsibility of the Ammunition Technical Officers (ATOs) of the Royal Army Ordnance Corps (RAOC)—before Northern Ireland they were dealing with IEDs in Aden, Cyprus, Hong Kong, Malaysia and other colonial troublespots.

The best way for an ATO to disarm an IED is by using his hands and his expertise. But it is often suicidal to approach an IED, so a method of dealing with these devices remotely was found—a vehicle was developed, of which several versions now exist, which allows the ATO to remain at a safe distance while he uses it to locate, identify and monitor a suspected bomb. If he decides that the object is too dangerous to be approached, he can attempt to disarm or destroy it by using various aids on the vehicle. One of the best-known of these vehicles is the Wheelbarrow: it has a TV camera and monitor to allow remote surveillance, lights to illuminate the target, a shotgun to break locks or windows, and a disrupter to render the firing mechanism of an explosive device harmless. Over the years, Wheelbarrow and other systems like it have saved the lives of many disposal experts.

There are two types of Wheelbarrow currently available: the first is the Mk7M, a radio-controlled and updated version of the successful Morfax Wheelbarrow Mk7 series; it was developed to operate in the demanding terrorist environment of Northern Ireland, and there are now over 500

'Keg-bombs'—beer kegs filled with explosive—in the boot of a car, the whole constituting a 'car bomb'. Note the Wheelbarrow remote control bomb disposal vehicle investigating the device

The Wheelbarrow Mk7 Super M

of them in service in more than fifty countries throughout the world. The vehicle is highly manoeuvrable and able to turn inside its own length. The older models of the vehicle were limited to an operative range of no more than 200 metres due to the umbilical cable, but the Mk7M can operate up to at least 500 metres, though the exact range is classified. Accessories include disrupters, shotgun, a drum of firing cable, x-ray equipment, car-towing hook and an electromechanical grab. A video colour camera with an auto-iris lens is fitted, which produces high definition pictures of the target on a colour monitor. The shotgun is used to gain access to a building or the boot of a car by shooting out the lock or shattering a window; the remote control grab or disrupter can then operate inside. The disrupter is loaded with a small high explosive charge and a metal projectile or bolt, and is electrically detonated; it can penetrate and break apart a bomb, without creating shock-loads high enough to initiate an explosion. If the bomb is too large to disarm with a disrupter, then a demolition charge may have to be laid in order to set off a controlled explosion; this is to separate at least part of the main explosive from the bomb. The charge can be put in place using the grab, and the firing cable unreeled as the vehicle returns to the operator.

The Wheelbarrow Mk8

The RO-VEH with some of its associated equipment,
including TV camera and monitor, shotgun, disrupters,
cable, etc

The second version of the Wheelbarrow is the Mk8, which has two main improvements: first, the operator now has an increased ability to adjust the vehicle's centre of gravity, which improves stability when negotiating stairways. Secondly, maximum speed has been raised to 1.67m/s (5.5ft/s)—this allows the vehicle to 'dash' to the target, although the 'creep' facility has been retained, allowing precise operation in the vicinity of the target.

Wheelbarrow is the best known and most widely used remotely operated bomb disposal vehicle, but it does have rivals. The A1 security RO-VEH (remotely operated vehicle) has many of the same facilities, but it is one third of the weight of a Mk8 Wheelbarrow. It achieves this by being smaller, but also by having no batteries on board—power is provided from the mains via an umbilical cable. This has obvious disadvantages as well as advantages; for instance, whilst it is easier

to climb stairs with a lower all-up weight, it is not so easy to tow heavy objects away from a target. RO-VEH is an excellent alternative vehicle which fulfils a slightly different need. It is probably true to say that it is more suitable for police work whilst the Wheelbarrow is better for military work.

There are other remotely controlled vehicles on the market; some of the best known are the Defence Systems Ltd 'Hadrian', a large vehicle with an impressive lifting capacity; the Kentree 'Hobo' which is marketed by Royal Ordnance; and the 'Hunter' remote control EOD vehicle, a joint Hunting Engineering and SAS Group venture.

The EOD vehicle that is worth special attention, however, is the 'Cyclops' developed by IMVEC Ltd. It is a remarkably compact little vehi-

The Cyclops EOD vehicle

cle only 750mm in length, 360mm wide (compared to Hadrian which is 1,470mm in length and 700mm wide) and weighing only 18kg (compared to Hadrian's 180kg); with attachments, the total vehicle weight is about 30kg, but this is still significantly lighter than the standard EOD vehicle (Wheelbarrow Mk8 weighs 240kg with an operational load). Cyclops' attachments include a ruggedised colour CCD TV camera and a disrupter. It is small enough to enter confined spaces in trains, buses and aircraft and, for instance, pass beneath chairs, but is still agile enough to negotiate kerbstones and staircases. It has already been ordered by the French navy to investigate unexploded bombs in the confined spaces of ships.

All remotely controlled EOD vehicles have essentially the same characteristics: they can deal with all types of terrain, including climbing stairs with gradients up to 45°. They allow remote surveillance via a video camera; they provide the means to force an entry with a shotgun; they can pre-empt an explosion by disrupting a bomb; they allow remote handling and lifting of devices with a grab, and the larger vehicles can tow away cars; they can place explosive charges; and they are controlled with a joystick. The picture on the CCTV monitor is, of course, the scene in front of the vehicle. If the vehicle has disappeared out of sight inside a house or around a corner, it is possible to steer it using just the picture, but control is easier if it can also be seen by the controller. Most important of all, these vehicles can, and have, saved lives. EOD men in Northern Ireland have been able to deal with hundreds of bombs remotely, some of which have exploded, destroying the EOD vehicle.

The Wheelbarrow and similar remote-control vehicles are normally operated by a team of three or four men. The British Army teams use converted Transit vans with a ramp to transport the Wheelbarrow. The Ammunition Technical Officer (ATO) will actually operate the Wheelbarrow when it is dealing with a bomb, though any other member of the team is qualified to set it up and drive it to the target. The van carries an array of bomb disposal equipment, including auxiliary equipment for Wheelbarrow. The accumulated expertise of an EOD team is enormous, and the remotely controlled EOD vehicle is the main weapon in their armoury.

SWAT Teams

In the United States the incidence of serious urban violence and large-scale crime has increased enormously, and in some circumstances conventional police forces have proved inadequate for the task, even though they are, as a matter of course, armed. Thus it became necessary in the 1970s for police forces to form special teams of police officers to tackle unusually dangerous situations involving armed criminals. These were termed 'special warfare teams' or SWAT teams. They were designed to tackle terrorist or hostage situations, sieges, drug 'busts', armed hijackers and any other potentially violent criminals. Their appearance and training is military in style and the individuals who are selected for SWAT teams are carefully chosen, often having to submit to a gruelling selection procedure. For those selected for a SWAT team, the régime is demanding. Teams are often on standby for twenty-four hours, and a high standard of fitness and of expertise in particular skills must be achieved and then maintained. In return, SWAT team members receive special rates of pay.

The composition of these teams varies from one police department to another. However, in general they approximate in size to that of a US Army squad, that is eight to ten men strong. There is a commander and a second-in-command, and the remainder of the team is often organised so it can divide into two sub-groups or fire teams. Often a target may need to be approached from two directions and only one team is available; so for tactical flexibility, each team must be capable of dividing. In addition to special expertise in standard skills such as accurate shooting, fieldcraft, first aid and surveillance techniques, SWAT team members require extra skills such as the ability to operate a tactical radio communications net; they need to be exceptionally physically fit, and able to undertake such hazardous duties as

A SWAT team member equipped with respirator, radio, tactical vest, grenades, handgun (in holster), powerful torch and Uzi sub-machine gun with image intensification device

rappelling, sniping, assaulting a house or aircraft occupied by armed gunmen, or rescuing a hostage. In so far as is possible, every member of a SWAT team should be interchangeable and be able to meet any of the demands put upon him. In the real world, where training is inadequate and other bureaucratic obstacles get in the way, some members of SWAT teams are given special responsibility for a role. For instance each team will have one member that has the highest skills in first aid—he is in effect a para-medic and the team 'doctor'. Inevitably one team member will be a particularly good shot; if there is a requirement to pick off a gunman at long range with a sniper rifle it will be his responsibility.

SWAT teams need special transport, and police departments have approached this problem in different ways. The majority supply their teams with specially adapted standard Chrysler or Ford vans in the tradition of the 'A Team'. These are provided with every facility including a special radio fit, seats for the team, racking for their equipment and the usual flashing lights and sirens. Some SWAT teams, however, have been equipped with armoured vehicles, such as the Cadillac Gage V-150 Commando, or the Cadillac Gage Commando Ranger. These are robust armoured vehicles which are proof against small arms fire and which are capable of mounting searchlights, tear gas grenade launchers, and even mounts for automatic weapons.

The exact role of SWAT teams is often misunderstood. The police in the United States have always been armed and are therefore equipped to deal with most situations involving armed criminals, but in the 1970s, with the increased threat of terrorism and of drug-related crime, police forces increasingly found themselves facing *organised* crime—crimes involving the use of firearms and carried out by a gang of criminals. The incidence of terrorism in the United States compared to Europe has been remarkably infrequent, but that of organised crime has shown a dramatic increase; much of this involves drug smuggling, protection rackets and armed robbery. There has also been a disturbing increase in mass killing carried out by a deranged gunman with apparently no logical motive; more than once this has involved school children being held hostage.

In all these situations, being able to call in a group of men who are specialists and, most important of all, trained as a team, stabilises the situation and offers some chance of a solution. In normal circumstances, policemen are trained to act either as an individual or, at best, as part of a team of two in a police car. It is this very independence which distinguishes the police officer from the military man who is used to operating in a hierarchical situation.

Thus SWAT teams, in many respects, are paramilitary in nature, more soldiers than they are policemen. In a hostage or siege situation, for instance, a building may need to be surrounded and watched closely for a long period of time. Snipers will need to watch windows and doors for a glimpse of their target. This requires sophisticated surveillance equipment capable of operating during both day and night. Finally, the building may have to be assaulted in a coordinated military-style operation. Hostages may have to be rescued. In such a situation, the normal police will hand over responsibility to the SWAT team, and will not become involved again until the assault phase is complete. When the shooting is over the SWAT team, who prefer to remain anonymous, will withdraw as soon as possible leaving the police to clear up the pieces.

SWAT teams require a massive inventory of sophisticated equipment to ensure that they are effective. They need high-powered weapons, hard body armour, abseiling equipment, loadbearing vests, tactical vests, surveillance equipment, grenade launchers for CS gas, special radios and more besides. Different SWAT teams are equipped with a variety of weapons. Among the most popular are the Austrian Steyr AUG-P, the Israeli Uzi submachine-gun, the Italian Luigi Franchi SPAS 15, the German Heckler and Koch MP5A3 and the German Mauser SP66 sniping rifle. Many SWAT teams are still equipped with the AR-15 or Armalite.

Body armour is clearly a 'must' in any assault situation. In both the USA and Europe there are many different alternatives on the market, from the most comprehensive and effective (and heavy!) versions with metal plate inserts which are capable of stopping high velocity rifle bullets, to the flimsiest so-called 'bullet-proof vests'—some of which are barely capable of stopping hand-gun shot. One of the best known body armour manufacturers is American Body Armour Equipment Inc. They produce a comprehensive range of body armour including facemasks, helmets, shields, tactical vests, lightweight military body armour and high coverage tactical armour. The model AK-47 lightweight military armour can, with its ceramic plates, defeat high velocity rifles: it weighs 2.65kg, and the 25cm × 30cm ceramic plate insert weighs another 3.63kg. Another US manufacturer of body armour, Second Chance Body Armour, produces a whole range of equipment, though the model most suitable for SWAT teams is the Hardcorps III. This is an all-purpose system offering 80.4cm² front and back protection from shell fragmentation, hand-guns and shotguns. There are three frontal inserts and an optional back insert protection against high velocity rounds. Hardcorps III without the plates weighs 2.47kg; with frontal plates, the total weight is 8.38kg.

All SWAT teams now wear so-called loadbearing vests or LBVs. These vests hold equipment close to the body so that it will not snag, yet will be readily available for instant deployment. Using the pouches, straps, packs and other containers on the vest not only keeps items close to hand, but also reduces any rattling noise during operations. A well-designed vest will distribute the weight of a load over a larger area, whereas standard military equipment tends to concentrate it on the shoulder where it is not easy to get at. An LBV can be worn over body armour; a tactical vest, on the other hand, incorporates body armour into the vest itself. An LBV can be with or without an integral rappelling harness, whereas SWAT team members using a tactical vest have to depend on a separate rappelling harness. LBVs carry magazines for pistols and submachine-guns, pouches for stun-, fragmentation or CS grenades, and space for a gas mask, as well as first-aid and radio equipment, plus all the standard items necessary for surviving in an operational environment for extended periods. All in all, SWAT teams are small, highly trained organisations which rely to a large extent on sophisticated equipment.

Their tactics are essentially military tactics: they are designed to allow the team to close with the enemy, if necessary eliminate him, and to do this with the minimum of casualties to the team or to any innocent bystanders. This adds up to a tall order. Most SWAT operations can be divided into three phases: the approach, the stake-out, and the assault.

The Approach

The approach involves using, if possible, covert means to approach a target. This involves using camouflage; it involves all the techniques of moving stealthily, of stalking, the leopard crawl, using covered approaches, reading the ground, and much more besides. If a covert approach is not possible, it will require the techniques of fire and movement—that is, always having some members of the team providing covering fire so as to keep enemy heads down in order to allow their teammates to make a dash closer to the target. This process is carried out alternately, so that the manoeuvre element becomes the fire element, thus allowing the second half of the team to edge closer to the target. The approach phase may involve the use of smoke grenades to cover a move forward, or the use of CS gas to distract the gunman. If CS gas is used, SWAT team members will have to put up with the additional inconvenience and discomfort of operating with gas masks. Part of the team providing the covering fire—perhaps one or two men—will be equipped with sniper rifles so they can pick off gunmen at long range if they show themselves for long enough.

The Stake-Out

The stake-out, or siege phase, may last a long time, days or even weeks. However, it is true to say that on past evidence, US SWAT team tactics usually seek quicker solutions than do British organisations, which tend to play a waiting game. Nonetheless, the teams are trained and equipped to seal off a gunman or gunmen for a protracted period; in particular this requires training in all the sophisticated techniques of surveillance, using image intensification and thermal imaging equipment for 24-hour, all-weather monitoring of the situation, as well as in using listening and 'bugging' equipment. SWAT team leaders must be trained in psychological warfare techniques, and must be able to negotiate with desperate men. During this stake-out they must collect, collate and interpret every scrap of information so that they build up an accurate intelligence picture. For this they will need some sort of operations centre, and SWAT team vehicles are equipped for this purpose.

The Assault

The assault phase requires the most complex and demanding training and tactics of all. Assuming negotiations have failed, an assault or rescue attempt will in the end have to be mounted. Sometimes there is no stake-out, and an assault has to be mounted instantly on arrival to prevent further loss of life, such as when a deranged gunman is already killing people. Either way, an assault of a building is a dangerous and tricky operation. It involves fire and manoeuvre, possibly wearing gas masks; it may involve scaling the

outside of a building using assault ladders, or rappelling down from a roof so as to enter a building through windows; it could even involve an approach to the roof of a building using helicopters. If it is an aircraft hijack it may involve blowing open an aircraft door under the cover of stun-grenades and CS gas. Each man will have a personal radio and earphones that allows constant contact between team members. When closing with the gunman, quick and accurate close range shooting will be required. And most difficult of all, SWAT team members must be taught to discriminate between innocent bystanders and armed criminals or terrorists. This can only be achieved by constant training.

SWAT teams have become an integral part of US police forces. They are available to operate in every state. They are part of the urban scene.

11 Weapons and Equipment of Urban Combat

Urban combat demands, for the most part, man-portable equipment. Standard infantry weapons and equipment are therefore much in evidence in this type of warfare. There are also, however, various items of equipment which are used solely or primarily in urban combat.

Sniper Rifles

One of these is the sniper rifle. Although snipers can, of course, operate in a rural environment, they are particularly effective in built-up areas—the density of the cover in a town will often mean that only fleeting glimpses are caught of the enemy. The standard infantry rifle is just not able to cope in such circumstances; moreover, telescopic sights are needed to see into a house. The typical sniper rifle is bolt-operated, is often fired with a bipod, has an adjustable butt and is fitted with a telescopic sight. It usually guarantees a 100 per cent first shot hit capability at all ranges up to 600 metres, and is sighted up to 900 metres. By contrast the average assault rifle is effective up to 300 metres and is sighted up to 600. Clearly these sorts of ranges are not usually required in urban

combat, but even at 100 metres it is difficult to hit a fleeting target in a built-up area. The British Army has found this to its cost in Northern Ireland, where the ratio of shots fired to hits achieved has been disappointingly low. But the sniper rifle is only suitable when firing from one strongpoint to another. Close-range combat requires an automatic weapon, so troops involved in house clearing will require an assault rifle or a submachine-gun.

Hand Grenades

Hand grenades are a pre-requisite of house clearing, and it is a sensible precaution to put one through a door or a window before entering a room. However, the supply of these weapons is usually limited, and there is certainly a limit as to

Sniper rifles are highly effective in an urban environment. This Parker Hale M85 Sniper rifle fitted with a SIMRAD KN250 night vision device on a Schmidt and Bender 6X42 telescope sight is one of the most effective sniper rifles available. Targets can be engaged accurately up to 1,000 metres

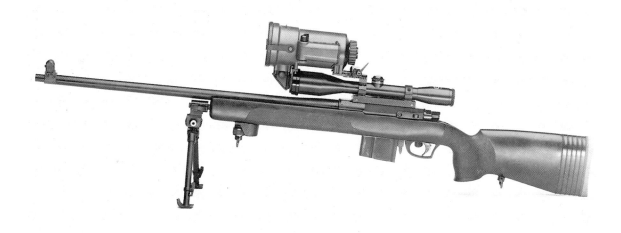

the number that can be carried. Contrary to the popular conception of hand grenades, their bulk and weight probably preclude an infantryman carrying more than about six. Extreme care must be taken when using grenades in close combat situations; they are as likely to cause casualties to friendly troops as to the enemy.

The most effective method of employing grenades is not to throw them, but to project them from a launcher. There are a variety of these systems available. They can be either single or multi-shot systems, are normally 40mm and are capable of launching an HE grenade some 350 metres with considerable accuracy. This gives an infantry assault team a stand-off capability: they can lob grenades through windows from the relative safety of another building and before they have to cross any open ground surrounding their objective. The US Army pioneered this category of weapon with the introduction of the 40mm M79 grenade launcher, which looks like a single-barrelled shotgun. The M79 fires grenades that are in effect small shells, the rifling in the barrel giving

the grenades considerable accuracy. Some practice is required to judge the range to the target accurately as the trajectory is high, although a trained soldier can put a grenade through a window at a range of 150 metres; however, the lethal area of the comparatively small grenades is restricted. The M79 was first introduced in the 1960s, but its bulk meant that the gunner could carry no other weapon; the US Army therefore replaced it with the M203 40mm grenade launcher some years ago. This fits under the barrel of an M16A1 assault rifle with its own trigger and sighting mechanism, allowing the grenadier to take part in the high velocity fire fight as well as firing grenades.

Anti-tank Weapons

Lightweight shoulder-launched anti-tank weapons are also particularly suitable for use in an urban environment, not only against tanks but against buildings, too. The IRA in Belfast and various Palestinian groups in the Lebanon, for instance, have used the ubiquitous Soviet-designed RPG-7 with some success against armoured vehicles. And there are many, far more sophisticated weapons of this type: one example is the British-designed LAW-80, which fires a 94mm shaped charge warhead capable of penetrating some 650mm of armour out to a range of 500 metres; it does this with unusual accuracy, which it owes primarily to its built-in spotting

The Singapore Industries CIS 40GL 40mm grenade launcher is one of several of its kind on the market. The CIS 40GL is a single shot, breech-loaded grenade launcher which is capable of projecting HE and other projectiles up to 400 metres. With practice, grenades can be fired through windows over a considerable range

rifle. But, although anti-tank systems such as LAW-80 can be used with considerable effect against buildings or bunkers, purpose-designed 'bunker-busters' are clearly that much more effective. Rifle grenades carry only a small amount of explosive and have a limited fragmentation radius, but Brunswick Defense in the USA have produced a rocket projectile designed to fit under a service rifle such as the M16 in exactly the same fashion as the M203. The 'Rifleman's assault weapon' (RAW) is a sphere, 140mm in diameter, which produces a 'squash-head' effect, enabling it to demolish walls and penetrate fieldworks or even armour plate. It is launched by firing a bullet in the normal manner; some of the cartridge gas leaving the muzzle is diverted to ignite the RAW's rocket motor via an ignition cap, and the flat trajectory of the projectile gives it an effective range of about 200m, although maximum range is well over 1,000m.

During the Falklands war it was revealed that Milan wire-guided anti-tank missiles made excellent weapons for destroying machine-gun or sniper positions. Both sides in the Vietnam war also used their anti-tank weapons as 'bunker busters'. The US Marines have recently adopted the 'shoulder-launched multi-purpose assault weapon' (SMAW); this is based on an Israeli 'bazooka' style anti-tank weapon, the B-300, which has too small a warhead to engage the latest generation of main battle tanks with any chance of success. The shaped charge warhead of SMAW will be used to attack pillboxes and field works from a range of up to 250m, which makes much more sense than extending guided anti-tank munitions or having to get close enough to put a grenade through a loophole.

Assault Equipment

Assault equipment is obviously a basic requirement of urban combat. Simple aluminium ladders are needed to gain entry into buildings if the enemy have sealed off entry via ground-floor doors and windows. An interesting new development, however, is that of air-powered mortars and projectile launchers which can launch grapnels and other similar assault equipment. Being powered entirely by compressed air, such systems do not generate any smoke, heat or flash signature, and can therefore launch, for example, metal grapnels with rope attached through windows or onto parapets relatively quietly, either to initiate covert operations, or when some degree of surprise is required in an overt assault. They have a range of about 100 metres.

If entry has to be forced at ground level—particularly in counter-insurgency situations—door-breakers or door-'busters' can be used. These equipments are placed against a door, locked into the door frame, and a hydraulic ram applied, its very precise force capable of opening the strongest door within a short space of time.

Flame Weapons

Flame weapons have been used with great effect in urban combat. They were widely used in the urban battles in Germany in 1945. Quite apart from its obvious effectiveness, the psychological effect of such a terrifying weapon system should not be discounted. One used by the British Army in 1945 was known as the Crocodile, in fact a Churchill Mark VII with a flame-gun fitted instead of one of the forward machine-guns. A two-wheeled armoured trailer, weighing 9 tons, was towed by the tank. This contained 400 gallons of special flame-burning fuel together with the nitrogen bottles which provided the necessary pressure to project the flame. This amount of fuel allowed approximately a hundred one-second shots, maximum range 120 yards. However, the system was relatively rudimentary and suffered from a number of disadvantages. In particular, tanks had to move within 60–80 yards of their target before engaging with flame, which depended on there being no enemy anti-tank weapons in the vicinity. Furthermore the trailers were only proof against small arms fire and tanks had to approach the target head-on—which depended on there being no additional or supporting enemy positions sited to a flank. Manoeuvrability was greatly restricted by the trailer. Also, before first going into action, thirty minutes had to be allowed for the raising of pressure, and it could not be sustained for long periods. Crocodiles could therefore only be used in conditions of superiority such as prevailed in Germany in 1945. They were a way of reducing casualties to Allied troops by eliminating the need to assault a stubborn strongpoint first with infantry.

1   *Sight*
2   *Carrying strap*
3   *Spotting rifle*
4   *Nose switch*
5   *Warhead*
6   *Warhead initiating train*
7   *Detonator*
8   *SAFU*
9   *Igniter*
10 *Non-electric initiation*
11 *HTPB rocket motor*
12 *Fins*

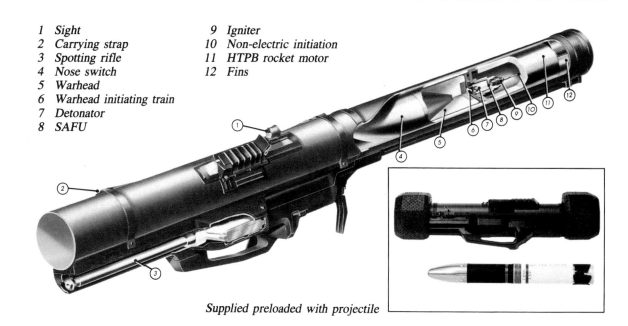

Supplied preloaded with projectile

The LAW 80 (Light Anti-Tank Weapon) is a lightweight, shoulder fired anti-tank weapon in service with the British Army. It is capable of destroying main battle tanks at ranges up to 500 metres. The 94mm shaped charge warhead penetrates armour in excess of 650mm thickness

If a SMAW is not available primitive demolition charges such as that illustrated here may be very efficient

Aluminium ladders being used to gain access to an upper storey

One such instance occurred early in the north-west European campaign, later in 1944. The following is an eye-witness account:

'The first attempt by a Canadian infantry battalion to capture May-sur-Orne was defeated with heavy casualties. That same afternoon the attack was renewed by the now seriously depleted battalion, but this time it had the additional support of four troops of Crocodiles. This second attempt was completely successful.

'C Coy and D Coy were combined—a total of about 60 men—and attacked on the *right* of the road with a troop of Crocodiles in support. To the *left* of the road, A Coy and B Coy advanced with two troops of Crocodiles. H-Hour was fixed for 1545 hours. The tank crews were assured that there were very few anti-tank guns in the neighbourhood, and throughout the action the Crocodiles were able to fight most aggressively. It was later discovered that the enemy positions had been very accurately pin-pointed by air photographs, and the defence overprint maps indicated every post, with one single exception, exactly as it was found on the ground.

'The Crocodiles moved forward, each tank closely followed by two sections of infantry, one just behind the trailer and the other 20 yards in rear. During the attack, any hedges or

The PSC Freyssinet Doorbreaker Model 2200 is a compressed air system with a working pressure of 10,000 psi, which can break down any door

ditches previously known to contain MGs or snipers were seared with flame. When the village was reached, the Crocodiles moved away from the road and one tank in each troop advanced on a line directly behind the houses while the other two tanks placed themselves so as to give covering fire.

'As each tank with the right two companies approached a house, it blew a hole in the building with its 75mm gun and squirted a jet of flame through the hole into the house, which then caught fire. The leading section directly behind the tank at once dashed for the doorway and cleared the house as quickly as possible. This procedure may sound dangerous, but the liquid, once ignited, is unlikely to cause damage to the attacker. He should, however, beware of getting in the path of the jet, because everything touched by the liquid catches fire. While the leading section was clearing the first house, the Crocodile moved on to the next house and again blew a hole in it and squirted in some flame; this time, the second section following the tank dashed for the doorway and cleared the house. Once the first section had finished its work, it came back to the tank—thus the two sections alternately cleared the buildings set fire to by the tank. The sections of the reserve platoon followed farther in rear to occupy the houses once they were cleared. The fires which were started blazed all night, and some of the buildings continued to burn all next day.'

The propulsion of projectiles by compressed air has many advantages. Flash is eliminated, noise levels are low and, perhaps most important of all, there is no ignition trail or evidence of smoke which can give away the location of the launcher. Plumett, a British Company, manufactures a range of air-powered mortars capable of projecting a variety of projectiles including line and grapnel carriers, smoke and flare canisters. Pictured here is the Plumett air launcher Type AL-52 which is capable of throwing a 3mm thickness rope about 100 metres

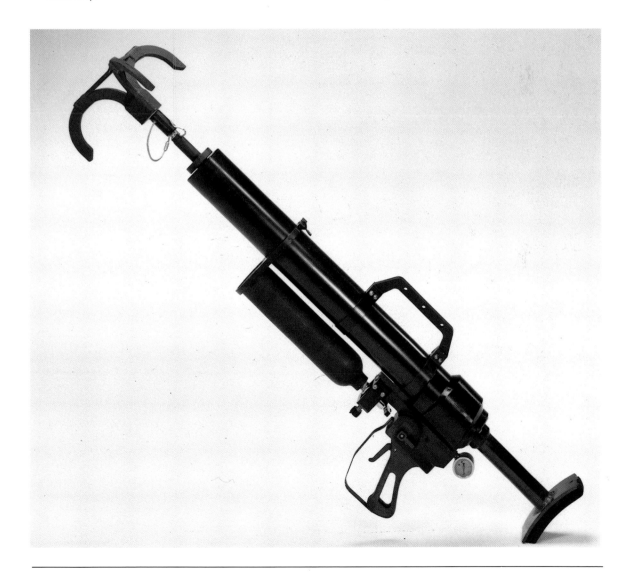

There have been no substantial developments in flame warfare since the end of World War II; this has been more a function of the political opprobrium its espousal would attract, rather than a reflection of its military utility.

Tanks

Tanks, as we have seen, can play a decisive role in urban combat, though they need careful protection by infantry. The classic role of tanks is to operate in open rolling country where they can move at speed and en masse from cover to cover; in such conditions they can keep their distance from any threatening short-range infantry anti-tank weapons and, with their superior firepower, engage them whilst still beyond their range. Clearly the urban environment presents conditions at the opposite end of the combat spectrum. During World War II tanks were, for the most part, only used in large numbers by an attacker with overwhelming superiority—for example in north-west Europe 1944–5. In less favourable conditions, the 'force multiplier' effect of urban terrain often makes tanks too vulnerable for use in close quarter fighting.

The greatest contemporary proponents of the use of tanks in urban warfare are the Israelis. Armour is a major ingredient in IDF urban warfare doctrine. In the Beirut battle, tanks were used to concentrate firepower on specific targets and to protect infantry. Israeli tanks often responded to hostile fire by shooting phosphorous shells into the buildings from which fire originated. Even when the target was not hit, the psychological effect of blast and noise played a part in causing Palestinian fighters to leave their positions. Tank fire was also used to breach walls.

Israeli military leaders were, however, much more cautious in their use of APCs. They were of the firm opinion that APCs are not properly equipped or sufficiently armoured for city fighting. Thus in Beirut they were only used for the transportation of equipment, supplies and casualties some distance behind the front line.

This Saracen APC has been fitted out for urban internal security operations. The 'outer skin' of mesh is designed to detonate shaped charge projectiles before they hit the armoured skin of the vehicle

Interestingly, after the war, the IDF modified a number of Centurion tanks by removing the turret and gun and by adding machine-guns and other infantry equipment. The resulting vehicle has since been used as an APC for road clearing and patrol duties. It is, of course, an immensely expensive solution to the problem; and probably one which only the IDF could espouse, faced—as they are—with the realities of war on their border.

Artillery

Artillery can also be a mixed blessing in an urban environment. During World War II it was employed in many urban battles to 'soften up' the opposition prior to an infantry assault, but seldom concurrently with close-quarter fighting in the target area. The resultant rubble in the streets often made progress extremely slow and created more problems than it solved. However, Israeli experience in Beirut sheds a different light on the matter, where the IDF used artillery to deliver high volumes of fire on point targets; for example, the Rafael David computerised fire control system was effective in coordinating the fire of widely dispersed artillery companies onto a single target, although there is evidence that profligate waste of artillery firepower was not unusual. However, the emphasis put on the direct laying of artillery pieces was innovative. SP artillery was brought close to the front line, particularly the US-manufactured 155mm M109, and targets were engaged using HE in the direct fire role. That is to say, gun barrels were aimed directly at targets which were engaged over open sights. This technique was used to breach walls too distant or too effectively covered by fire to reach by more conventional means.

Mortars

Mortars can be effective in urban terrain, though they are more traditionally, and usefully, employed to engage troops in the open. But because of their high trajectory they can sometimes be the only weapons system which can attack troops behind a high obstacle such as a high-rise building or an elevated urban motorway. It is often particularly difficult to engage armoured vehicles hidden in and amongst buildings.

A new generation of mortars may be set to change this, however. The British Merlin system under development is a 'smart' 81mm (3.2in) mortar bomb with an active millimetric seeker in the nose, a guidance computer and control surfaces which deploy in flight, and a top-attack anti-armour warhead. The US Army has a development requirement for a similar system called GAMP (guided anti-armour mortar projectile). Range will be about 8km (5 miles) and guidance of the IR type, with a top-attack self-forging projectile doing the damage to enemy armoured formations. The French have followed another route in the production of a 120mm round effective against lightly armoured vehicles within a radius of 14m. All such 'smart' systems are particularly relevant in an urban context where point targets are numerous.

LLAD Guns

Low-level air defence (LLAD) guns have been used *in extremis* in the past against ground targets. They were, for instance, used by the Argentinian garrison in defence of the settlements of Goose Green and Darwin during the Falklands war in 1982, against the British infantry as it was closing in on the base. So effective was this fire that the Paras were unable to advance without the likelihood of sustaining unacceptable casualties. Thus a Harrier strike was called in to eliminate this particular threat. The Palestinian forces used LLAD guns against Israeli ground targets during the Beirut confrontation, and indeed they remained in constant use by both Moslem and Christian organisations in Beirut until late 1990.

Surveillance and Other Equipment

Surveillance equipment is particularly relevant in the urban environment. By definition an urban area is exceptionally dense; targets are therefore more difficult to locate, and greater resources need to be invested in identifying them. Buildings more than two storeys high pose particular problems for target acquisition. As a guide, the 'dead' space behind a building is generally five times the height of the building for normal indirect fire and only half the height for high angle indirect fire. Thus mortars are particularly suited to urban operations, although their penetrative power is limited. One way of reducing the surveillance and target acquisition problem is to use aerial vehicles.

Helicopters are likely to be too vulnerable to ground fire, and fixed-wing aircraft fly too fast to be able to locate pinpoint targets in a dense urban area—or even if they can locate a target, it is likely to take too long for the information to be transmitted to those on the ground who need to know the answer within minutes. The solution to this dilemma is the 'remotely piloted vehicle', or RPV. These are small, unmanned, fixed-wing air vehicles mounting a downward facing TV camera which is capable of transmitting 'real time', ie simultaneous TV pictures to a ground station. The Israelis used RPVs in the Beirut battle to provide real time intelligence on PLO movements in the city. They were also used to provide immediate feedback on the success of air strikes and other attacks.

The selection of projectile and fuse type is also dictated by the urban environment—for example, the effects of projectiles with proximity fuses are severely reduced by structures. High explosive with a delayed fuse mechanism is the optimum method of achieving maximum penetration of buildings. On the other hand, the effect of shaped charge or heat projectiles is much reduced if chicken wire or some other form of outer skin is added to vehicles or buildings. This technique is used by the British Army in Northern Ireland around permanent bases, where a protective barrier of wire netting is erected around the perimeter; this has the effect of harmlessly detonating shaped charge projectiles—such as those fired by the RPG-7 shoulder-fired anti-tank weapon—*before* they can impact upon structures inside the perimeter. A similar 'skirt' can be erected around parts of an armoured vehicle in a counter-insurgency situation. But the real breakthrough in projectile design in urban warfare is so-called 'smart' weaponry. Now, laser-guided and other precision-guided munitions permit the attack of targets with minimum damage to adjacent buildings.

Perhaps the most difficult problem facing the infantryman in the heat of the urban battle is identifying from where enemy fire is emanating. In and amongst buildings—and amid the fear and confusion of battle—rifle shots echo and re-echo around the urban landscape, and it is virtually impossible to locate the source of fire. However,

the British Army has come up with a novel solution to this problem, deploying a system called 'Claribel' in the urban areas of Northern Ireland. Small radar-heads are located in the four corners of a vehicle, with consequent coverage of 360°. When a shot is fired at the vehicle, the round is detected by one of the radar-heads and the direction from which it has been fired is indicated by a flashing light on a clock-face display inside the vehicle; the crew therefore looks in the right direction to try and locate their target. A man-portable version was also produced, with the radar-heads attached to a soldier's helmet and the direction of attack indicated on a wrist-watch-type display. Although this was less successful, the idea is an ingenious one, and an example of how a terrorist threat can accelerate the pace of weapons research and development.

Training

Realistic training for urban combat is the key to effectiveness. After the experience of Calais in 1940, the British Army dedicated large areas of bomb-damaged towns to training for urban combat. Apparently the lesson has not been lost on those who allocate priorities for contemporary training facilities. Large FIBUA complexes have been built or are planned for the British Army in both the United Kingdom and in Germany. One such village has recently been completed on Salisbury Plain training area. More interestingly, a facility has existed for some years which prepares soldiers for a variety of urban contingencies in Northern Ireland. This is a range, rather than a training area, in the sense that individual soldiers or pairs of soldiers are dispatched along a course of incidents armed with their service rifle. This is fitted into a sub-calibre device which allows a .22 round to be fired at any targets which may appear, rather than a high velocity 5.56mm round. Incidents are designed to introduce an element of choice: suddenly a very realistic target simulating an armed terrorist will run fleetingly across an open space—but at the same time, another mechanism will interpose a dummy representing a mother pushing a pram with an infant in it. The soldier must curb his instinct to open fire, in the interests of safety. Further down-range, a curtain in an upper storey window parts and a gunman

(Above) A house used for urban combat training on a range in Berlin

(Opposite, above) A Land Rover fitted with 'Claribel' detectors

(Opposite, below) All equipment used in urban combat must be camouflaged appropriately. Here a Land Rover of the British Berlin Brigade is painted in the rectangular patterns of urban camouflage

fires: this time there is nothing to stop the soldier returning fire. The range is nothing more than a film set, but it is a highly effective training aid.

Urban combat is equipment-intensive. Development of different categories of weapons and equipment in an attempt to achieve some sort of advantage in the urban battle has been continuous, though not spectacular. The only category of weapon which has been put aside since the end of World War II is the flame-thrower. It is, of course, particulary effective in an urban environment—though armies do not appear to have had the same qualms about air-delivered napalm or artillery-delivered white phosphorous. Perhaps flame has not been used because the circumstances in which it would have been useful have not occurred. Clearly such a weapon could only be used in general war; the political price for use in low-intensity operations would not be worth any limited military advantage that might be gained.

Recent history would seem to suggest that, despite the improvement in East/West relations, the incidence of low-intensity conflagrations is unlikely to subside, and that many of these are likely to occur—in part, at least—in an urban environment. Armies would be wise to continue to train and arm for urban combat.

12   Northern Ireland: A Case History

Of all the post-war examples of urban warfare, Northern Ireland has been the most persistent. It must be emphasised that Northern Ireland is not just an urban conflict—indeed for the most part it is a counter-terrorist campaign fought in a rural environment; but its highest profile aspects have usually been in either Belfast or Londonderry. Also it should be made clear from the outset that the many crowd control and riot situations faced by the British Army in Northern Ireland, whether inter-sectarian or purely anti-security force in nature, cannot by any stretch of the imagination be termed urban combat. But there have been many occasions where gun battles between the security forces and the IRA have taken place in an urban environment in Northern Ireland. It is these situations, and the tactics adopted by the security forces to frustrate the IRA in Belfast and Londonderry, that can properly be analysed in the context of urban warfare.

Northern Ireland

Ever since the beginning of the campaign of terrorist violence undertaken by the IRA in Northern Ireland in 1969, patrolling by the security forces has been the main method of dominating the cities and seizing the initiative from the terrorists. Indeed, patrolling in Northern Ireland has two main purposes: domination of the ground, so as to deny the enemy freedom of movement; and to get to know the area intimately in order to build up a detailed knowledge of it and its inhabitants.

From 1969 to 1971 patrolling was reactive rather than preventive. Battalions had some difficulty in even keeping up with the pace of events; they were seldom able to take the initiative. As the years passed, however, patrolling maps were updated, and the sheer volume of intelligence on the inhabitants of the battalion or company area of responsibility provided such a degree of back-

Belfast

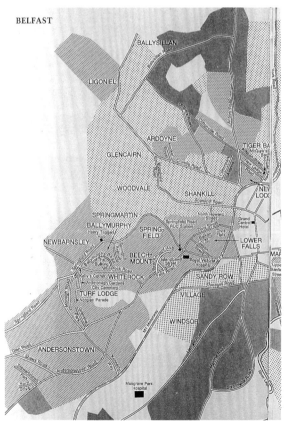

ground information that patrol commanders were able to put a name to most of the faces they passed in the street. This is even more the case in the early 1990s, since the army has had over twenty years in which to build up a mosaic of information.

In Belfast or Londonderry, battalions were each given a particular area of responsibility. In the early 1970s the West Belfast Battalion was responsible for the Falls, Whiterock, Springfield Road, Collins, Beechmount and Ballymurphy areas (all Catholic) as well as the Highfield, Springmartin, Glencairn and Village areas (all Protestant). In 1974 the author was responsible, as a company commander, for part of this area, namely the Collins, part of the Springfield Road as well as the Highfield, Springmartin and Glencairn estates. The total population of this area was approximately 25,000 and included two large factories, a wood-yard, a home for mentally handicapped children, a slice of Blackmountain itself—a large hill on the edge of Belfast—and the interfaces between the Catholic Ballymurphy and the Protestant Highfield estates and the Protestant Springmartin and Catholic Collins. The requirement was to dominate the area and to learn as much about it as possible in four-and-a-half months.

So as not to have to start from scratch, each battalion sends an advance party to Northern Ireland some weeks before the main body arrives; normally this would consist of the commanding officer, the company commanders, and platoon and section commanders who will then tramp the ground with patrols from the battalion they are relieving. Thus by the time the battalion's second-in-command, the company seconds-in-command and the platoon sergeants bring over the bulk of the men, the commanders have made themselves

(Opposite, above) This photograph shows an urban combat training range somewhere in the UK, used to prepare troops for tours in Northern Ireland. Targets appear in windows and soldiers are able to engage them using a .22 sub calibre device fitted in their rifles

(Opposite, below) A foot patrol keeping alert in an urban area in Northern Ireland. Whilst some soldiers move, others lurk in doorways covering them

familiar with the ground. Furthermore, the battalion intelligence officer precedes the advance party by several weeks, so that he can assimilate the accumulated knowledge of the intelligence officer of the battalion being relieved. This system of relief took some years to develop and has now been refined to a drill. During the early years, however, battalions were rushed out at little or no notice as both the government and the military merely reacted to events.

The whole system is designed to maximise the knowledge of a battalion or company area for the short six-month roulement tour. These constraints do not apply to the same extent to a two-year tour as part of the garrison.

Whatever the length of tour, proper preparation and training for a Northern Ireland tour is, of course, absolutely vital. In the early years training was haphazard. For a start, few were absolutely certain what to train for. Now there is a well-oiled training machine, which puts every battalion through a standard Northern Ireland training package; this includes intensive patrolling, either urban or rural depending upon battalion location, riot-control techniques, shooting at fleeting targets, first aid, powers of and procedures for arrest, orders for opening fire, IRA bomb and weapon recognition and capabilities, IRA techniques, capabilities and organisation, as well as training in the use of various items of internal security equipment. Soldiers now go to Northern Ireland well prepared and trained. There are, in fact, a number of locations where urban combat ranges have been constructed. These are rather like film sets, with house façades on either side of the streets looking incredibly realistic but with nothing behind the façade. Soldiers patrol these 'streets', though with sub-calibre devices fitted in their rifles so that high velocity bullets are not winging all over the countryside. Instead, a .22 round is fired at targets which appear in windows or down an alleyway. The test is not to fire at everything that moves, but to discriminate between innocent occurrences and terrorist threats. These urban ranges are excellent training aids.

A vehicle patrol leaving a security force base. It will be accompanied by another vehicle

It is important that patrolling is co-ordinated as a haphazard system is unlikely to produce results, nor will it dominate the area. Conversely a predictable and repetitive patrolling plan, however well co-ordinated, can be used by the enemy to mount ambushes. A company commander will therefore try to achieve a balance between these two sometimes conflicting requirements.

As a general rule, there will always be someone on the ground twenty-four hours a day. This was certainly the case at the height of the troubles in the early and mid-1970s, though it may not always be the case today when the army's role is more reactive. That presence need not be obvious. It could be, and often is during the early hours of the morning, a static presence in the form of a covert observation post (OP); a well-sited OP can often dominate large parts of a company area. Indeed the OP need not be covert; it may be on top of an obvious block of flats and its presence may be well known—though if this is the case it will, of course, have to be well guarded and defended.

It is patrolling, however, on foot or in vehicles, that actually dominates an area—the physical presence of soldiers prevents the enemy from preparing or planning an illegal activity. Having said this, the IRA would argue that the presence of soldiers on the streets is provocative and the catalyst for their terrorist activities. But the rule of law cannot be maintained without regular visits from those upholding it. Patrols must be seen regularly on the streets to give confidence to the local population. In some respects vehicle patrols have a high profile and are easier and safer to mount, particularly if an armoured vehicle is used; though the best method of securing information and of really getting to know an area is on foot. The majority of patrols in an urban area must always be on foot. The company commander must therefore plan a matrix of vehicle and foot patrols which covers the entire company area twenty-four hours a day in an irregular and unpredictable pattern, in such a way that no patrol is ever left unsupported by another patrol. Into this pattern he will work the odd static OP. Thus all patrols are mutually supported, and indeed some are 'multiple' from the outset; that is to say, two or more patrols will cover a grid of side streets working in parallel in a co-ordinated manner and in radio

contact. In this way they will discourage an ambush by possibly cutting off escape routes.

All patrols will be given a task or tasks by the company commander. These may include:
1. Setting up a vehicle check-point (VCP) on a particular road for a limited period.
2. Updating house occupancy in a particular road or street.
3. Visiting a factory at opening or closing times to prevent sectarian attacks.
4. A security visit to a sub-post office or garage, both favourite targets for robberies.
5. Visiting families which may have been the subject of sectarian threats.
6. Looking for suitable sites for future OPs.
7. Searching derelict buildings and wasteland for hidden arms or explosives.
8. Monitoring a parade or funeral.

There are many other tasks, and the plan must be constantly updated and checked to ensure that the security forces are a step ahead of the enemy. Above all, patrol commanders must be debriefed by the company commander after each patrol—only in that way can the intelligence 'jigsaw' be kept up to date. The most apparently insignificant snippet may be of value at a later date, for instance the fact that a new family has moved into a particular house.

Before a spell of duty the patrol commander, his briefing complete, leads his four-man unit to the sandbagged bunker by the entrance to the base. Pointing their rifles into the bunker they cock their weapons and then run zigzagging out of the gate. They are instantly 'on patrol', away from the comparative safety of the company base where, even if they can be mortared, they at least cannot be shot at.

An Ammunition Technical Officer (ATO), or 'bomb disposal expert' in the vernacular, makes a manual approach to a suspected 'keg bomb' in the boot of a car. Sometimes a manual approach is necessary, as it is not possible to manoeuvre a robot disposal device sufficiently close. Obviously, this is a much more dangerous option

Typical of the mean streets of Belfast, these terraced houses are covered in Loyalist murals depicting 'King Billy'—King William of Orange on a white charger

As the patrol commander leads his men into the area he has been told to investigate, he will be conscious of several things: perhaps most importantly of all he will be looking into every window and doorway, every street corner and hedgerow for a possible telltale sign of an ambush—something glinting in the sun, an open window, a curtain moving, something that could be construed as a signal given by boys to a waiting gunman or bomber. He will also be responsible for keeping his patrol together, watching each man and ensuring he is carrying out his allotted task. On top of all this he will be navigating: however familiar he and his patrol are with the ground, he does have to be aware all the time of precisely where he is—in the event of a contact he must be able to report his location instantly over the radio. Next, as patrol commander he will be responsible for communicating over his pocketphone with Company HQ and with other patrols supporting him or working with him. Lastly and most important of all, he will be carrying out whatever the patrol task is: as patrol commander *he* will have to fill out a written patrol report after the patrol; and he it is who will have to carry out identity checks, or check the occupants of vehicles at a VCP. In short, the pressure will be on the JNCO all the time. The chances are that he is a young 23-year-old corporal who has been in the army for five years; this may be his first, perhaps only his second Northern Ireland tour. The responsibility, by any yardstick, is unusually great.

In the early 1970s the emphasis was very much on overt patrolling; today it is public knowledge that there is a greater emphasis on covert operations, the aim to be preventive rather than reactive. In urban areas, observation posts are set up in derelict houses and on rooftops. In one incident in 1974 in the author's own area of responsibility in the Collins district of Belfast, a platoon commander had set up an OP in a derelict semi-detached house, with a good area of observation and field of fire looking down the Springfield Road. After it had been in position for two days, a shot rang out from what sounded like the other side of the wall against which the three men in the OP were leaning. After extricating themselves from their cramped positions, they were just in time to see a gunman disappearing around a street

corner 150 yards away. It transpired that he had taken up a position very quietly several hours before, in the other of the two semi-detached houses (which was lived in), and had fired at a patrol on the Springfield Road—which fortunately he missed. This is illustrative of the sort of cat-and-mouse game that is played in the backstreets of Belfast.

The Royal Engineers are responsible for providing expert search teams in Northern Ireland, but they cannot possibly cover the whole province. For this reason each battalion sends soldiers, usually from the Assault Pioneer Platoon, to the School of Engineering at Chatham to attend a 'search techniques' course prior to a Northern Ireland tour. These men are taught search techniques in both urban and rural environments; they are also taught how to disarm booby traps, since as team members, they are obviously more likely than most to be confronted with these while searching for arms, explosives and ammunition.

In urban areas clearly there are many hiding-places. Any attic or loose floorboard, maybe the inside of a false wall can accommodate an illegal weapons cache. Nor is it possible to search every house on an estate in a systematic manner. First, it would cause such a storm of protest that it would be counter-productive in PR terms. Second, it would require so many soldiers to cordon and monitor an estate in order to prevent material being smuggled out at one end while the other end was being searched, that it would render the prospect a non-starter. Third, it would probably be more effective to use some, or all of the soldiers, in some other way. Thus searches are normally only undertaken on receipt of specific intelligence. Then permission has to be sought (and is not always granted) from Brigade Headquarters, before a search of a private house is sanctioned. Unless in 'hot pursuit' after a shooting incident, even a battalion commander may not authorise entry into, or a search of, a private residence. If, as a result of specific information, permission is granted for the search of a house, the householder will be informed that his house is going to be searched and he will be invited to accompany the search-party during their search. If it is necessary to cause any damage such as lifting floorboards, the damage is always repaired by special teams. The householder will be asked to sign a form saying that no damage has been caused or, if it has, that it has been or will be repaired. Obviously searching houses causes resentment, which is why it is only undertaken if there is positive intelligence that weapons are being concealed.

Search teams are equipped and trained to do their job as quickly and efficiently as they can, and to cause as little damage as possible. Apart from a range of mundane items such as jemmies, spades and screw-drivers, they are equipped with metal detectors, explosives 'sniffers', inspection mirrors and fluorescent hand-lamps. Most explosives 'sniffers' will indicate the presence of gelignite, dynamite, nitro-glycerine, nitro-benzine and some other types of explosive. If explosives *are* located, then it is the task of specialist bomb disposal teams to come and deal with them.

The skill of bomb disposal has developed rapidly since 1971, which is when the IRA started to use the bomb widely as a weapon. The British Army already had 'explosive ordnance disposal'

(EOD) personnel before the present emergency in Northern Ireland, mostly employed in disposing of World War II German bombs. These teams, found from the Royal Army Ordnance Corps and Royal Engineers, have various disposal aids available to them. In particular, remotely-controlled vehicles have been developed, capable of carrying and operating a variety of equipment necessary for the locating and disposal of dangerous objects. These vehicles enable the ammunition technical officer (ATO) to remain at a safe distance while he locates, identifies by TV camera, and monitors a suspected bomb. If he decides that the object is too dangerous to be approached, he can attempt to disarm or destroy it, using various aids on the vehicle.

In many instances a manual approach is

The Wheelbarrow bomb disposal vehicle makes an approach to a suspect car bomb. The Wheelbarrow has investigated hundreds of suspected bombs over the years in Northern Ireland, and is used worldwide

necessary, either to prevent blast damage or because forensic evidence is required. In such a case the ATO will wear an EOD suit which is designed to give some protection against fragments, blast and flame during the disarming of small, improvised explosive devices (IEDs). It will also provide a measure of protection at greater range against larger 'devices'. If a manual approach is made, the ATO will carry an inspection set consisting of light probes, extension rods, mirrors, magnets, lock viewers and hooks and line. All these components are made of nonferrous metals. Should a remote operation be possible, the 'Wheelbarrow remote bomb disposal equipment' is used. The Wheelbarrow is a concept which first saw operational service in 1972. The vehicle is powered by two reversible electric motors running off two inboard 12V batteries; it has a range of 110 yards, limited only by the 18-way control cables through which commands are passed. Its mean endurance is two hours. Wheelbarrow, which has been progressively developed over the years, can now attack virtually any IED. And if the ATO runs out of time, it is the vehicle which is damaged rather than a man being killed. The latest versions of Wheelbarrow are extremely sophisticated systems.

Northern Ireland urban operations have always required careful co-ordination. Different patrols need to inform each other of their location so that one does not accidentally confront another, or even fire on the other in error; contacts have to be reported to HQs, the police kept informed, helicopters requested, suspects' details checked against records, bomb disposal teams called in. All this, and much more besides, must be done on the radio.

In an urban IS situation it is not convenient or efficient to use conventional radio equipment. In 1969, when the British Army first deployed on the streets of Belfast and Londonderry, they brought with them their Larkspur A41 radio sets. It was fortunate that at that juncture the IRA threat had not developed sufficiently for instant communications to be a prerequisite for at least adequate operations in an urban environment, because the

An infantryman must move cautiously along a street. He watches every window, every doorway, every alleyway for a telltale sign of ambush

A41 was a disaster in and around buildings. Often it was necessary to fly a Sioux helicopter as an airborne relay station to ensure efficient communications. And if the helicopter was shot at or ran out of fuel, soldiers on the ground often had to resort to the GPO telephone system!

Conventional military radio equipment is, by and large, too heavy, bulky and complex for use in the streets. Most important, VHF radios are designed to operate most efficiently when there is a reasonable line of sight between radio stations. Buildings get in the way. It therefore became necessary in the early 1970s to introduce small so-called Pocketphone radios which operate in the VHF and Ultra High Frequency (UHF) bands via permanent rebroadcast stations. A purchase was made of hand-held Pocketphone UHF-type PF5-UH single-channel sets operating between 405 and 440 MHz or 440 and 470 MHz. These had a 15V rechargeable NiCd battery and gave up to twenty-four hours' operation. In a situation where troops are operating from static bases it is important that transmissions be scrambled or in cipher. Thus over the years the army has been issued with off-the-shelf 'civilian' equipment which has provided efficient and simple communications between patrols on the ground and Company and Battalion Headquarters. In addition, the army has the ability to detect and locate illicit transmitters.

Most armoured personnel carriers (APCs) today are tracked, and as we have already seen, are not suited to IS operations for a number of reasons: they are difficult and expensive to operate and maintain, they are noisy, they cause damage to roads and, most of all, they are classed as 'tanks' by the layman. Use of tanks in an IS situation is often politically unacceptable, and conjures up visions of Soviet tanks in the streets of Prague. Therefore most armoured IS vehicles are 4 × 4 wheeled vehicles, which afford protection from small arms fire up to 7.62mm ball. They are provided with observation blocks and firing ports. Vulnerable points on the vehicle, such as the fuel tank and the radiator, must be given special protection, and the vehicle must be designed to allow rapid exit from and entry into the vehicle by the crew and passengers. These vehicles can be fitted with a variety of armaments including water cannon, tear-gas launchers and machine-guns. Some can even be electrified to prevent terrorists climbing on to the vehicle. In addition to armoured vehicles, there are several other types of vehicle that are commonly used in urban IS situations, including water-cannon vehicles, which may or may not be armoured; and conventional 'soft-skin' vehicles, which have been covered in a form of appliqué lightweight armour as protection against blast and low-velocity rounds, and armoured bulldozers for the removal of barricades.

A photograph of the GKN Sankey AT-104 'Flying Pig'. It has been nicknamed the 'Flying Pig' because of the unfolding anti-missile barriers on either side of the vehicle. These are used to protect soldiers from stones, bottles and any other missiles thrown by demonstrators

In Northern Ireland the British Army still employs the GKN Sankey AT-104 IS vehicle commonly known as the 'Pig' as its standard APC for use in an urban environment. The 'Pig' has been adapted in many different ways: for example the 'Flying Pig' can unfold large fenders from each flank so that if parked in the middle of the road, it can block most of the road off, and afford protection against missiles thrown by rioters.

Urban patrolling in Northern Ireland is also carried out in Land Rovers. The British Army was forced to develop GRP and Macralon armour for standard Land Rovers in an attempt to provide some protection against blast, fire and acid bombs, and low-velocity small arms fire. The RUC are equipped with the Hotspur Armoured Land Rover, the grey painted police vehicle so familiar

on the streets of Belfast. It is equipped with 'add-on' 4.76mm Hotspur steel, and the windscreen, side and rear windows are of 29mm laminated glass. Another Land Rover variant is the Shorland MK.3 armoured car, employed by the UDR.

These, then, are some of the techniques and tactics used in urban IS operations in Northern Ireland; they have been developed over a period of twenty years and are tried and tested. The overriding principle which governs the British Army urban warfare tactics in Belfast and Londonderry is that of 'minimum force', and this is something which is particularly British. Most nations' armed forces use the technique of maximum force in order to solve any given situation, the idea being that an overt and massive use of force will terrify any urban terrorist enemy into submission. These tactics have been traditionally employed by the Soviet Army in such situations as the Hungarian uprising in 1956 and the invasion of Czechoslovakia in 1968; by the Indian and Pakistani authorities in the suppression of violence; by the

Although the doctrine of 'minimum force' has been consistently applied in Northern Ireland, there have been occasions when this has not always been the case. In situations where crowds are hurling both abuse and missiles, soldiers are sometimes goaded into reaction

Communist Chinese in Lhasa in Tibet in 1988; and of course in Tiananmen Square in 1989—and even by the French and German police in riot situations in Europe. It can be argued that the use of maximum force is extremely efficient; even if there are casualties in the short term, it usually prevents a recurrence of the situation and so saves lives in the long term.

The British Army has always relied, however, on the technique of employing the minimum force necessary to solve any given situation. Hence tracked vehicles have never (except for operation Motorman in 1972) been employed in Belfast or Londonderry; automatic weapons may not be employed, and soldiers may only fire single-aimed shots; permission has to be granted at a fairly high level to fire even rubber bullets or tear-gas rounds; and each soldier carries on his person the so-called 'yellow card' which lays down precisely under what conditions and in what circumstances a soldier may open fire. The exact content of the card is classified information, but in the broadest terms a soldier may only open fire with live ammunition if he believes that his life or the life of another person is in imminent danger. Even then he must, if practicable, shout a warning that he is going to open fire. Thus violence meted out by the authorities is strictly controlled, and only sufficient 'doses' of armed force are used to solve a given problem. The doctrine of 'minimum force' is in the best traditions of British justice and fair play, and ensures that—in so far as is possible in a violent situation—life is not unnecessarily endangered. It gives the maximum opportunity for those who are breaking the law to reconsider their course of action. Whilst perhaps not the most militarily efficient doctrine, it is both humane and constructive. It avoids creating martyrs for a cause and, in the international context, puts the British authorities in a favourable light. Whether or not, however, maximum force in 1969–70 would have avoided a further twenty years of bloodshed, we shall never know.

The urban landscapes of Northern Ireland have seen many different kinds of confrontations between the British Army and the IRA in the last twenty-one years. Two stories, one a success for the army and one a disaster—which could have been even worse had it not been for the bravery of a Green Jacket Corporal—will serve to illustrate the nature of urban warfare in this troubled land.

The first story concerns the 3rd Battalion the Royal Green Jackets, a distinguished battalion from an equally distinguished regiment that has a record of outstanding success in Northern Ireland. The 3 RGJ area of responsibility in 1973 conveniently covered most of the territory terrorised by the 2nd Provisional 'Battalion' of the IRA. This organisation, until shortly before the arrival of 3 RGJ, had numbered about eight officers and thirty or more volunteers, but attrition by successive British regiments over the preceding months had steadily reduced its numbers to a few high-ranking officers and low-grade juvenile volunteers. Such a process of attrition had, of course,

to be achieved by exclusively legal means. There were still enough snipers, gunmen and bombers to make life unpleasant, however, and 3 RGJ's first incident occurred on the afternoon of 28 July when a foot patrol was fired at by more than one gunman from a range of about a hundred yards. A burst of automatic low-velocity fire and several high-velocity shots missed the patrol, but went through the glass entrance door of some pensioners' flats.

A few days later, in New Barnsley, an S Company patrol from the Anti-Tank Platoon noticed a young woman with a somewhat curious gait. On closer examination her discomfort was found to be caused by a .303in sniper rifle complete with telescopic sight. Two days later the familiar marzipan-like smell of explosive led another patrol to search a derelict house where they uncovered one rifle, a revolver, assorted ammunition and a quantity of explosives. This seemed to open the floodgates, so that by the last week in August various finds netted eighteen weapons including a Soviet-manufactured RPG-7 rocket launcher and 2,600 rounds of assorted ammunition.

Then, on the last day of August, 3 RGJ shot the infamous PIRA terrorist, Jim Bryson. It is worth examining this incident in some detail. Bryson, who came from a fiercely Republican family, had acquired a reputation with the RUC for bullying and brawling during his youth. He grew into a squat, broad-shouldered, evil-looking man, who readily joined the IRA when the troubles started in 1969, some would say in order to indulge his homicidal tendencies. During the escalation of the insurrection in Belfast in 1971, Bryson developed into a cunning and ruthless terrorist. He operated mainly in the Ballymurphy area where his crude leadership and shooting exploits made him into something of a cult hero. In June 1972 he took command of the Ballymurphy Provisional Company, ruling it and the people of the Ballymurphy by a system of terror which demanded and perforce received universal obedience. He was also extremely active himself, and is believed to have murdered a number of soldiers and policemen with his Armalite rifle fitted with a telescopic sight. He was arrested in November 1972, but escaped from the Crumlin Road courthouse in March 1973. He fled to Eire, but was

A rioter throwing a petrol bomb at security forces in Northern Ireland. If he is endangering life, he can be shot

asked by the Provisional Brigade staff to return to the Ballymurphy in August to help redress the balance against the Official IRA whose influence had been growing in the area. He immediately started to terrorise the local Official IRA men, who decided that Bryson would have to be executed. Such was the degree of mutual distrust among the two wings of the IRA in the Ballymurphy on 31 August.

On the morning of the same day, an S Company corporal and a rifleman climbed stealthily into the attic of a flat directly overlooking a circle of open ground surrounded by council houses, known as 'the Bullring'. Missing roof tiles afforded them a good view of the Bullring and the roads leading off it. This was one of a pattern of three or four observation posts which S Company had set up for collecting tactical intelligence and maintaining general surveillance over the area. Soldiers would stay in the OPs for several days on end whereupon, if the position had not been compromised, they would be relieved by another team.

At 1830 hours that evening it was the corporal's turn on duty; he had seen nothing of interest all day and was bored. Subconsciously he noticed an olive green Hillman Hunter approaching the Bullring. Suddenly, to his astonishment, he saw three rifles sticking out of the car's windows before it drove off out of sight. He reported on his radio to Company HQ. Moments later another OP reported on the radio net that the car was continuing to cruise around the area, now followed by a red van. It transpired that Bryson, accompanied by three notorious PIRA terrorists—Paddy Mulvenna, ex-adjutant and commander of the Ballymurphy Company, 'Bimbo' O'Rawe and Frank Duffy—were driving around the Ballymurphy, partly to demonstrate their disregard for the army and partly to humiliate the Officials.

The Hillman and the red van reappeared in the

Bullring, drove slowly around it, and then stopped at a junction some fifty yards beyond. The occupants got out and Bryson began to direct them to ambush positions. The corporal carefully moved one of the roof tiles to one side so as to get a better view and to give himself a fire position. However, he inadvertently dislodged a tile, which clattered down to the ground and alerted the ambush party, one of whom fired in the general direction of the OP. The corporal immediately fired four rounds, although he could scarcely aim from his cramped position. He was forced to pull his rifle in when it developed a stoppage, and this enabled Bryson and his gang to escape. When the corporal looked again, the road was empty. Their position now compromised anyway, the corporal and the rifleman set about enlarging the hole by kicking more tiles out. The corporal stuck his head out to try and get a better view—and withdrew it sharply as two rounds hit the roof. Glimpsing a gunman, he fired three rapid rounds, but missed.

Thinking that the gunmen were now making good their escape, both soldiers hurriedly prepared to leave the OP. As they were doing so, the corporal was amazed to see the Hillman returning to the same junction. Ironically, Bryson had become confused by the same problem that faced so many British soldiers in Belfast, that of determining where the fire had come from. In built-up areas it is virtually impossible to tell from which direction a shot is fired, since the 'crack and thump' of a high velocity round echoes and re-echoes off the walls of the tightly-knit Belfast streets. Bryson, thinking he was driving *into* danger, had thrown his car into a wild U-turn. As they came back into view, passed the OP and were driving away from it, the corporal fired at the accelerating Hillman. The first round hit O'Rawe in the shoulder and catapulted him from the back seat into the front of the car. Then a 7.62mm round entered the back of Bryson's neck. As he slumped forward the car careered into the small front garden of 99 Ballymurphy Road. The two soldiers watched the car crash some two hundred yards away, and jumped down from the OP in the attic to the flat below where they took up fire positions to cover the car.

Mulvenna was the first to recover; he flung open the door of the car and rolled to the ground

from where he engaged the OP soldiers with his Armalite. Duffy also began to fire from the back of the car with an MI carbine. Mulvenna then decided to make a run for it. As he did so the corporal fired three shots, two of which hit, and Mulvenna died instantly. The next to go was Bimbo O'Rawe who, though wounded, was still clutching a Garand rifle as he ran towards the door of 99 Ballymurphy Road. Again the corporal fired three shots, hitting O'Rawe as he pitched forward inside the house. He then turned his attention to Duffy who was firing wildly as he sprinted away down the road. The corporal fired but missed.

When S Company patrols arrived they found Mulvenna dead, Bryson unconscious and O'Rawe badly wounded. Bryson died three weeks later. In the follow-up, thirteen rifles and pistols and large quantities of ammunition and explosives were found. The perseverence, alertness and good shooting of S Company had rid Ulster of three heartless murderers. In all, that August six gunmen were killed, bringing the total number of terrorists put out of action in one way or another to 2,265, including 195 Protestants. In Belfast, the three Provisional battalions ceased to exist. In their place the Provisionals created small 'Active Service Units' (ASUs) based on the Communist 'cell' system, whereby members would be known only to others in the same unit, and whose commanders would be directly responsible to the Belfast commander, Ivor Bell. Ironically, the Provisionals were convinced that Bryson had been eliminated by the Officials; of course, no one had seen from where the shots had come. Soldiers only appeared on the scene after the event. It suited the army to perpetuate the myth.

The remainder of the 3 RGJ tour was equally eventful. There were more shootings and several bomb explosions, including one car bomb of some 450lb of explosive which slightly injured two Green Jacket soldiers. An A Company patrol wounded a sniper who habitually shot at sentries in McRory Park base from the City Cemetery; blood trails confirmed the hit. Finally, on 28 November, the battalion returned to England. Compared to many of the tours of 1971–2, it had been quiet, but it was very typical of the countless tours that have now been undertaken by every major unit in the British Army: four to six months of continuous activity of eighteen-hour days, of constant patrolling and unrelenting tension; nights spent in cramped accommodation in bunk beds, or perhaps days and nights on end in an OP in a derelict building; the very real danger from bullet and bomb; the hatred of bigots, but also the gratitude of the vast majority. Four months of this and the strain would show.

The second story concerns the 2nd Battalion the Royal Green Jackets, an equally distinguished and famous battalion of the same regiment. They, too, have had many successful tours in Northern Ireland. Their tour in West Belfast from December 1981 to March 1982 was relatively quiet, as indeed was the city as a whole, although they did manage to find thirty-five illegal weapons, 3,615 rounds of ammunition, twenty-seven grenades and 9lb of explosive. One of the surprising features of this particular tour was the extent of ordinary crime; in the 2 RGJ area of responsibility there were more than seventy armed robberies in the space of four months. Kneecappings and the occasional inter-sectarian murder occurred with depressing regularity. Then tragically, only two days before the battalion was due to return to England, the 2 RGJ tour was marred when a mobile patrol was fired on close to the RUC Springfield Road police station—three riflemen were killed in a highly organised PIRA ambush involving an M-60 machine-gun. There was also a bomb left at the firing-point, designed to catch the follow-up forces, but mercifully it failed to detonate.

Corporal William Lindfield was in command of a mobile patrol which was taking an RAF Sergeant from Brigade HQ to the C Company base at North Howard Street Mill. The patrol got into its two Macralon Land Rovers inside the Springfield Road police station and drove out of the back gates. They turned right, and right again down Crocus Street back towards the Springfield Road. When the leading Land Rover, driven by Cpl Lindfield, was 50 yards down Crocus Street, all hell was let loose as automatic fire was brought to bear on it from very close range. Also in the Land Rover were Rifleman Daniel Holland, Rifleman Mark Mullen, a lance corporal from the Coldstream Guards (who were taking over from 2

RGJ) and the RAF sergeant. Cpl Lindfield, realising that his vehicle had been hit, accelerated out of the killing zone as fast as he could. He drove across the Springfield Road into a side-street where he stopped. Rifleman Holland had received gunshot wounds in the head and was unconscious, the RAF sergeant had also been shot in the head and was bleeding badly, although still conscious. The Guards L/Cpl, although he had been hit in the head by a ricochet, was able to look after the other wounded men. Cpl Lindfield rushed back across the Springfield Road with Rifleman Mullen to where the other Land Rover was standing in the killing zone.

Its driver, L/Cpl Darral Harwood, having seen the other Land Rover hit, had endeavoured to get himself and his crew out of the vehicle before they reached the killing zone. He had managed to fall out of the driver's door, dropping his rifle in the process, but his companions had been unable to get out so quickly. Rifleman Anthony Rapley had been hit in the back of the head and had died instantly; Rifleman Malakos had received gunshot wounds in the stomach, neck and jaw. Another Guardsman was unscathed but was in a state of shock, as also by now was Rifleman Mullen who had attempted to assist Rapley, only to find that he was dead. L/Cpl Harwood dragged Rapley's body behind a car, leaving Cpl Lindfield to run under fire to the door of the house whence the enemy fire was coming. By now, reinforcements had arrived from the nearby Springfield Road police station; they had heard the firing and were able to prevent Lindfield from going any farther. For his gallantry Lindfield was awarded the Military Medal. The citation reads:

'. . . In all his actions were beyond reproach and throughout he displayed a coolness, courage and professionalism of the very highest order and which was an outstanding example and inspiration to all round him. His actions won open praise from police and civilians who witnessed him and his grip of the situation was mainly instrumental in re-establishing order in a very confused and shocked situation with multiple casualties.

'His prompt and correct actions at the moment of contact, with no consideration for his own personal safety, undoubtedly saved further serious casualties, and his gallantry and aggressive reaction to enemy fire were in the highest traditions of the service. His professionalism, leadership and gallantry both during and after the incident have been exemplary and present the strongest case for official recognition with a high award.'

In the incident, Rifleman Rapley was killed instantly, Rifleman Malakos died on the way to hospital, and Rifleman Holland died on the operating-table. The RAF sergeant recovered from his wound.

These are two incidents of urban combat from Northern Ireland involving only small numbers of men—one a stunning operational success, the other a disaster for those who so tragically died and for their families.

Northern Ireland has also seen large-scale urban operations, and the most famous was undoubtedly Operation 'Motorman' in 1972. For months, areas of Londonderry had been turned into 'No Go' areas for the security forces. They were controlled by the IRA, and the Rule of Law had effectively ceased in this particular part of the United Kingdom. It was not that the security forces were unable to enter these areas—they could have done so at any time. But it was judged politically too risky to allow the army into the Bogside for fear of the casualties that might ensue.

Enough was enough. At 0400 hours on 31 July, Operation 'Motorman' was launched, to enter and re-occupy the 'No Go' areas. A total of 21,000 troops were concentrated in the Province for the operation, and the intention to re-occupy the 'No Go' areas was broadcast ahead of the event in order to give the IRA the opportunity to leave the areas peacefully and so avoid bloodshed. As a result Operation 'Motorman' met virtually no resistance.

One of the battalions involved was 1 RGJ, which had been stationed in the pretty Hanoverian town of Celle in West Germany. In the prelude to Operation 'Motorman', Battalion

The aftermath of an IRA bomb in Belfast. This sort of devastation is a regular feature of Belfast life

The absurd situation in Londonderry before 'Operation Motorman', with IRA vigilantes openly guarding parts of the city. It was scenes like this that the operation stopped

Headquarters, A, C, and Support Companies had returned from training at Vogelsang in the Moselle region on 12 July, and B Company from training in Hohenfels in southern Germany on 23 July. At this stage most members of the battalion had plans for leave during August or September, and indeed by late July the commanding officer, Lieutenant Colonel Bob Pascoe, was heading south for Spain, complete with family, car and caravan. One major and two captains were sunning themselves in the South of France, to say nothing of nearly eighty members of the battalion on leave or on courses in England. Then it happened. At thirty minutes past midnight on Wednesday 26 July, the second-in-command was called to the telephone by Brigade Headquarters and told that the battalion was off to Northern Ireland, the advance party within twenty-four hours and the main body on Friday and Saturday. By 0300 hours a list of eighty-five names and addresses was telephoned through to the duty officer at the depot in Winchester, who telephoned local police stations to ask them to contact the 1st Battalion men in their area with the message to return to Germany immediately. Within twenty-four hours, all but one were back in Celle.

The area of responsibility for 1 RGJ in Operation 'Motorman' was the Andersonstown area of Belfast, where the IRA had held almost complete control since earlier in the year. All but the main roads were barricaded, IRA check-points controlled who entered the area, and few people paid their rent, electricity or gas bills. Stolen cars were used quite freely, and many cars were untaxed or uninsured. Normal services such as refuse collection and road sweeping had not been done properly for up to two years, street lights had been shot out in gun battles and never replaced, and gunmen roamed the streets with almost complete freedom, meting out their own justice. The police had not dared enter for fear of their lives for many months, and intimidation and fear were so rife that a member of the security forces could be wounded in the street and nobody would so much as look at him.

By the time such 'No Go' areas had become firmly established, it was too late to do anything about them without risking enormous civilian casualties. In accordance with the British tradition of minimum force to solve any given situation, the government of the day bided its time and only acted when it judged the least civilian casualties would be caused. And it was largely in order to keep these to a minimum that the public announcement of the pending operation was made. As already mentioned, there was virtually no opposition—as soon as 1 RGJ entered the area they took over two school buildings and set about dominating the area.

The Marines were even able to incorporate a beach landing in Operation 'Motorman'. Four landing craft from HMS *Fearless* made the 25-mile passage up the Foyle estuary to land Royal Engineer Centurion bulldozer tanks to remove the barricades in Londonderry. Escorted by the minesweeper HMS *Gavington*, the craft reached the landing beach just before midnight, coming ashore over trackway which had been laid quietly by hand so as not to attract undue attention. When the bulldozers had cleared the barriers in the Creggan and the Bogside, they returned to the beach and were taken off soon after dawn. In the course of this operation, a gunman and a petrol bomber were shot dead. In the Bogside, arms finds included an anti-tank rifle and a 0.30in Browning machine-gun.

In Belfast, eleven battalions moved into likely trouble spots in a coordinated operation: 42 Commando into Ligoniel, 2 Para into the Ballymurphy and Whiterock areas, 1 Prince of Wales Own and 2 Royal Regiment of Fusiliers into Andersonstown, 1 Light Infantry into the Ardoyne, 40 Commando into New Lodge, 1 Welsh Guards into the City Centre and the Markets, 3 Royal Anglian into the Beechmount, and 1 King's into the remaining parts of the Falls Road area; 19 Field Regiment Royal Artillery and the Life Guards were also involved in the operation. Total arms finds for Operation 'Motorman' totalled 32 weapons, over 1,000 rounds of ammunition, 450lb of explosives and 27 bombs.

Operation 'Motorman' was an outstanding success. For months the Provisionals ruled the Creggan and Bogside areas of Londonderry, as well as parts of Belfast. The situation had been allowed to persist only because intelligence had indicated that an attempt to take the 'No Go' areas would have resulted in massive civilian casualties. The political decision was therefore taken not to interfere, and although perhaps hard for a soldier to understand, this was in line with the British policy of minimum force, a policy which had successfully solved so many of the terrorist situations in the years of withdrawal from the Empire. For example in Aden in 1967, the GOC, General Philip Tower, did not re-occupy Crater, the heart of Aden town, which had been taken over by elements of the South Arabian Army (SAA) and Aden Police, because to do so successfully would have required considerable force, including probably the 76mm guns of the Saladin armoured cars. Instead he waited, and the Argylls chose their moment to re-enter Crater stealthily by night when most of the rebels were sleeping, and so virtually without opposition. In many respects this bears comparison with the re-occupation of the 'No Go' areas in Londonderry and Belfast. Yet such a policy cannot be popular with men who are trained to find quick and efficient solutions to problems, and it says a great deal for the British Army that it has always been able to understand the wider implications of a situation and to react accordingly. This cannot be said of all armies, even of some of those of our Western allies.

Republican propaganda in Londonderry marking the confines of the 'No Go' area smashed by Operation Motorman in 1972

The war in Belfast today is of a very different nature. Government policy requires is that the army should take a lower profile. Ever since 'the Way Ahead' policy was instituted in the early 1980s, the police—the Royal Ulster Constabulary (RUC)—have taken the lead in all security matters; routine patrolling is primarily, though not entirely, the business of the RUC and the Ulster Defence Regiment (UDR). By and large the army's role is again reactive. There is a much greater emphasis on covert operations—whilst the detail of these is clearly classified, it is no secret

that a number of military patrols and observation duties are undertaken in plain clothes and in civilian vehicles. It is also no secret that SAS men do operate in Northern Ireland, though normally they are only deployed as a result of specific intelligence of an intended shooting or bombing—in such circumstances they are able to place themselves in situ ahead of the event and catch the terrorists red-handed. There have been several instances of this happening during the 1980s. Perhaps the most notable occasion was not in Belfast or Londonderry, but in Gibraltar in 1988, when intelligence was received that an IRA 'Active Service Unit' planned to detonate a car bomb during the changing of the guard ceremony outside the Governor's residence—the Convent—in

The spot in Gibraltar where the three Irish terrorists were shot dead by SAS operatives

the narrow and crowded streets of Gibraltar. Three known IRA terrorists were carefully watched, from their arrival in Spain to their placing of a car next to where the band and the rear Guard would form up prior to marching onto the square in front of the Convent for the ceremony. It was not known whether this car contained explosives, but in view of the identity of the three terrorists and the nature of the intelligence which the authorities had received, it was a fair assumption that the car was a car bomb. In the event it contained no explosives, but was certainly parked there to reserve a space for the car bomb which would have been brought into the colony the following day prior to the ceremony. Three SAS men were tasked to apprehend the three ter-

rorists before they left British territory. As they approached the border with Spain on foot, the three SAS men shouted a warning. The terrorists made as if to pull hand-guns, whereupon they were shot dead by the SAS men.

This was a classic and very successful covert operation in an urban environment. It is also the sort of urban warfare which British security forces find themselves involved in to an increasing degree as we enter the 1990s. It calls for a high degree of training and ability, and very few soldiers have the potential to undertake such operations successfully. The failure rate in the selection procedure for the SAS and for those servicemen who are chosen to undertake covert operations in Northern Ireland is extremely high. This is hardly surprising, bearing in mind the nature of the duties these men are expected to undertake. They are required to operate undercover, often in and amongst the enemy in Republican areas, and in small groups; and at the same time they must be ready to go into action at a moment's notice. They must be able to shoot accurately and, more important, selectively. Much of what they are expected to do will take place in public places where innocent bystanders are going about their daily business. This sort of operation is far removed from urban warfare in the classic sense: nevertheless, urban warfare it is.

13   Some Conclusions—and the Future

Urban warfare is largely a twentieth-century phenomenon. This is hardly surprising, since before the Industrial Revolution cities were much smaller, they were less numerous, and they tended to be more densely constructed so that any rapid movement—let alone combat—was virtually impossible. It was therefore inevitable that, as urbanisation became increasingly widespread, combat would take place more and more within city boundaries. But it was not just a matter of geography that determined the increasing incidence of urban combat. In both low-intensity situations and in conventional conflict cities became strategically more significant: first, because a guerilla war initiated in an urban area has, from the point of view of the insurgent, a higher public relations' value; and second, because cities tend to have a greater concentration of high value targets such as industrial capacity, transport facilities, communications assets and the like.

Thus some of the bloodiest battles of World War II took place in urban areas: Stalingrad, Berlin, Caen, the Ruhr, Monte Cassino, Tobruk, Ortona, Goch. Nor has the incidence of urban conflict diminished very much since the end of World War II, mainly because revolutionary organisations have recognised the publicity value of this form of warfare, but also because the Soviets and the Chinese have recognised that an ideologically unsound population is best brought to heel by imposing law and order in the capital first—Prague, Budapest, Kabul and Beijing are relatively recent examples of this phenomenon. Also, great power interventions have often been aimed at cities, most recently Suez, Grenada and Panama City.

Certainly the technology of urban warfare has advanced dramatically in recent years. In particular the advent of so-called 'smart' weapons has meant that artillery and air-delivered weapons can be used in more precise and surgical strikes in an urban environment. Port Stanley in the Falkland Islands can hardly be categorised as an urban area; however, during the last days of the Anglo-Argentinian war over those islands in 1982 there was one fascinating example of urban precision bombing against a single 155mm gun located amidst civilian housing on the outskirts of Port Stanley. Clearly it was important not to hit civilian buildings or to cause casualties to civilians, both of which were less than 100 metres away from the Argentinian gun. A single Harrier aircraft of the ground attack variety was scrambled to destroy the lone gun which was threatening to disrupt the British advance into Port Stanley. The strike was entirely successful, and was particularly important because it showed the use of laser-guided munitions in modern warfare, demonstrating that these systems *are* effective in an urban scenario, or at least in a scenario where it may be necessary to destroy a single military target comparable in size to an urban strongpoint. There were, of course, many such examples of precision bombing during the Gulf War. In particular, it was possible to be highly selective in pinpointing military targets in Baghdad.

Hand-held weapons are also becoming more accurate with the introduction of various 'bunker-busters' which provide the infantryman with a greater 'stand-off' capability. However, this innovation is still unlikely to eliminate altogether the need for the infantryman to occupy buildings in urban combat, and therefore to close with the enemy and to engage in room-to-room and even hand-to-hand combat. In other words, despite the introduction of more effective weaponry, urban combat will remain an essentially primitive form of warfare, certainly for the forseeable future.

The trend towards greater urbanisation in Europe and in Third World countries is set to continue. Most forecasts suggest that although the building boom of the post-war years will not be repeated, there will still be a gradual increase in urbanisation in Europe. Moreover the recent cataclysmic revolutions in central and eastern Europe are likely in the long run to provide the necessary stimulus to the economies of these newly emergent democracies. Increasing rates of urbanisation are an inevitable result. Similarly, a gradual improvement in the economies of some Third World nations—though certainly not all, probably not even most—could have the same effect elsewhere. Thus the statistical chance of conflict occurring in an urban scenario will certainly remain high, if not increase.

Building trends could have some effect on the future nature of urban warfare. The tendency to build cheap structures designed to last perhaps only 25–40 years in the 1950s and 1960s appears to have ended—at least in Europe—in the 1980s.

Building standards have improved and are returning to traditional designs, so theoretically at least there will be a greater availability of potential strongpoints. At the same time many European cities, particularly in Germany, are being opened up with the construction of wider highways and boulevards. There are longer views and wider vistas. Such a tendency would, of course, make it more difficult to block streets and erect obstacles; it will mean that there may be room for the use of armoured vehicles in the traditional manner, that is as *mobile* gun platforms rather than static pillboxes.

Most military experts, ignoring recent developments in Europe for the moment, seem to think that investing in additional FIBUA/MOUT training facilities is a sensible precaution. The British Army, for one, has very recently constructed a large new FIBUA training facility in the United Kingdom on Salisbury Plain Training Area. This has been built to represent a typical German vil-

A staff officer plans the techniques of urban warfare for the future using a model as an aid

lage, though no doubt the decision to build this facility was taken before the remarkable events of late 1989. Nevertheless if warfare—worldwide—is equally or more likely to occur in an urban area, this was probably a worthwhile investment.

To attempt to forecast accurately the likely future incidence of urban warfare is of course virtually impossible. However, looking at current trends, one or two educated guesses can be made. Clearly the Soviet Union is entering a period of considerable instability, particuarly in some of the southern republics. Similarly the Balkan States show signs of increasing instability. The Middle East situation shows little sign of immediate improvement. A form of urban combat—albeit unarmed for one combatant—continues with the 'Intifada'.

In late 1990 and throughout January and February of 1991, a small but determined Kuwaiti resistance movement made regular attacks on Iraqi military convoys and high-ranking officers in Kuwait city. Photographic evidence of such attacks reached the Western press even during the war. Clearly small numbers of Kuwaiti fighters took maximum advantage of the urban environment in Kuwait city to launch guerilla attacks on the Iraqi occupation forces. Retribution was swift and massive as the Iraqi army plundered the city

and occupied private dwellings, murdering many innocent citizens. But the Iraqi invaders were unable to prevent completely a small band of determined Kuwaitis from undertaking an urban guerilla offensive.

Terrorist organisations are likely to continue to capitalise on the advantages of the urban environment to further their objectives. The Northern Ireland problem seems set to continue for the forseeable future. Terrorism, if not exactly urban combat, is likely to continue to occur in Central America, South America, in the Indian sub-continent, in the Middle East and even in European cities—ETA bombings in Madrid and other Spanish cities being just one example. There are unfinished revolutions in the Balkans; Beijing has already seen one uprising; the government in Manila has put down a number of attempted army revolts; anything could happen in the cities of South Africa; the Sri Lankan problem continues to fester; Korea remains divided; fighting in El Salvador seems likely to continue.

Peace appears to have broken out in Europe as the Cold War comes to a sudden and unexpected end. But as we have seen, conflict in a number of arenas seems set to continue: urban combat, a child of the twentieth century, shows no sign of dying as the century draws to a close.

Notes

Chapter 2

1 Airey Neave, *The Flames of Calais* (Hodder and Stoughton, 1972) pp131–2.
2 Ibid p231.
3 Ibid p232.
4 Quoted in Geoffrey Jukes, *Stalingrad, The Turning Point* (MacDonald & Co, 1968) pp20–1.
5 Ibid p63.
6 Ibid p106.

Chapter 3

1 The 101st Airborne Division, a light infantry (paratroop) division, with some additional artillery, forty tanks and a tank destroyer battalion, should have been no match for two Panzer divisions and a Volks Grenadier division.
2 Notes on town clearing by commanding officer of 1st Battalion The Gordon Highlanders, February 1945, p1, a battalion in 153 (H) Brigade.
3 Ibid p2.
4 Its widespread use in the British Army was new, but flame warfare had been introduced by the Germans in 1915 during World War I.

Index